PLANT RESISTANCE TO INSECTS

A Fundamental Approach

C. Michael Smith

Department of Entomology
Louisiana State University
Baton Rouge, Louisiana

Department of Plant, Soil and
Entomological Sciences
University of Idaho
Moscow, Idaho

A WILEY-INTERSCIENCE PUBLICATION

JOHN WILEY & SONS

New York / Chichester / Brisbane / Toronto / Singapore

Library of Congress Cataloging-in-Publication Data:

Smith, C. Michael (Charles Michael)
Plant resistance to insects/C. Michael Smith.
p. cm.

"A Wiley-Interscience publication."
Bibliography: p.
Includes indexes.
1. Plants—Insect resistance. I. Title.

SB933.2.S64 1989 89-30629
632'.7—dc 19 CIP

ISBN 0-471-84938-3

Printed in the United States of America

10 9 8 7 6 5 4 3 2 1

To my family, Rita, Segen, and Sonder,
and to my parents, Glen and Joyce,
for all the faith and confidence they placed in me.

Preface

The theory, study, and use of plant resistance to insects has matured greatly since the publication of R. H. Painter's classic *Plant Resistance to Insects* in 1951. Interest in this area of insect pest management is actively growing in virtually all areas of world crop production today. New research information in areas such as molecular biology, predictive modeling of plant and insect genetics, and induced allelochemical resistance has emerged since the 1979 publication of *Breeding Plants Resistant to Insects* by F. G. Maxwell and P. R. Jennings and *Principles of Host Plant Resistance to Insect Pests* by N. Panda.

Plant Resistance to Insects: A Fundamental Approach has been prepared as a textbook for students and researchers in entomology and related plant protection disciplines. The extensive reference lists that have been prepared to document the points of interest in each chapter will also be of value to teachers and researchers in the areas of plant–insect interaction, insect pest management, and plant breeding.

The book has been arranged into three sections; each poses a question to the reader. The first section, "What is plant resistance to insects?," deals with the history of the subject and each of the three functional categories of plant resistance to insects: antixenosis, antibiosis, and plant tolerance. The second section asks the reader "How is plant resistance to insects obtained?" Chapters in this section deal with how potential sources of insect–resistant plant material are located, the techniques used to evaluate and incorporate them into desirable sources, and the factors that can affect the integrity of these evaluations. In the final section, "How can plant resistance to insects be utilized?," readers are offered overviews of the current and future uses of insect-resistant plant material in various crop insect pest management systems.

I am grateful to my colleagues at both the Louisiana State University Department of Entomology and the University of Idaho Department of Plant, Soil and Entomological Sciences for their encouragement and support during the development of this book. Special thanks are expressed to E. A. Heinrichs, Sharron Quisenberry, and Tom Sparks for their critical reviews of early versions of Chapters 8 and 10. I appreciate the efforts of Deborah Woolf, who typed several revisions of the original manuscript, and express my thanks to Maureen Riley and Minha Chu for their help in reference compilation and literature retrival. I also thank Mary Thrasher, graphics artist, Louisiana Agricultural Experiment Station, for preparation of line drawings and Colwell Cook and Li Dao Ke for assistance in preparation of photographic prints for publication. I acknowledge the support of the administration of the Louisiana State University Agricultural Center for granting me sabbatical leave, during which I completed a working draft of the book.

I am especially grateful to two true practitioners of plant resistance to insects, Frank Davis and Charlie Brim, for their inspiration and guidance, which helped me plan and develop this book. Finally, I thank the many students who have enrolled in my classes during the past ten years for their eagerness to learn and for their constructive criticism of my concepts of plant resistance to insects.

C. MICHAEL SMITH

Moscow, Idaho

May 1989

Contents

Section III. How Can Plant Resistance to Insects Be Utilized?

SECTION I

What Is Plant Resistance to Insects?

1 Introduction

1.1 USES OF PLANT RESISTANCE TO INSECTS

The cultivation of plants that are resistant to insects is a plant protection technique that has been in use for several hundred years. Before the domestication of plants for agricultural purposes, those susceptible to insects died before they could produce seed or before their damaged seeds could germinate. Thus resistant plants survived subject to the laws of adaptation and natural selection. With the advent of crop plant domestication and the use of rudimentary agricultural practices, farmers selected the seeds to use for future crops. Though there are few written accounts, it is probable that early agricultural systems in Africa, the Americas, and Asia selected and utilized plants resistant to insect pests, for these systems developed production practices based on different crop species and, within species, selected different strains and races of crops. Dicke (1972) describes the selection and development of mulberry genotypes by Oriental agriculturists that yielded high silkworm production and associated fine quality silk. In this instance, the plants' susceptibility to insects, instead of resistance to insects, was actually promoted.

During the eighteenth and early nineteenth centuries, insect-resistant cultivars of wheat and apples were first developed and cultivated in the United States. As early as 1788, early maturing wheat cultivars were grown in the United States to avoid infestation by the Hessian fly, *Mayetiola destructor* (Say) (Chapman 1826). A few years later, Havens (1792) identified resistance to the Hessian fly in the wheat cultivar 'Underhill' in New York. Lindley (1831) made recommendations for the cultivation of the apple cultivars 'Winter Majetin' and 'Siberian Bitter-Sweet,' because of their resistance to the woolly apple aphid, *Eriosoma lanigerum* (Hausmann). In the mid-nineteenth century, plant resistance to insects played an important role in Franco-American relations. American grape cultivars were found to be resistant to the grape phylloxera, *Phylloxera vittifolae* (Fitch), whereas French varieties were extremely susceptible. The world-famous French wine industry was destroyed by 1884, resulting in a $2 million loss. However, by subsequent grafting of resistant rootstocks from the United States to French scions, the industry recovered.

Fewer than 100 reports of plant resistance to insects were published in the United States during the nineteenth and early twentieth centuries. In one of the earliest comprehensive reviews of plant resistance to insects, Snelling (1941) identified 163 publications dealing with plant resistance to insects in the United States from 1931 until 1940. Since then, numerous reviews have chronicled the progress and accomplishments of scientists conducting research on plant resistance to insects (Beck 1965, Green & Hedin 1986, Harris 1980, Hedin 1978, 1983, Maxwell et al. 1972, Painter 1958).

The first book on the subject of plant resistance to insects, *Plant Resistance to Insect Pests,* was written by Painter (1951), who is considered by many individuals as the founder of research on plant resistance to insects in the United States. In the late 1970s intensified research activities in plant resistance prompted publication of several additional texts on the subject, including those of Russell (1978), *Plant Breeding for Pest and Disease Resistance*, Maxwell and Jennings (1980), *Breeding Plants Resistant to Insects*, and Panda (1979), *Principles of Host-Plant Resistance to Insects.*

Research involving the development and use of insect-resistant crop cultivars has led to significant crop improvements in the major food producing areas of the world in the past 40 years. For this reason, insect resistance occupies an important position in many national crop protection programs. One of the most spectacular successes of the use of plant resistance to insects occurred during the "Green Revolution" in tropical Asia during the 1960s, when high-yielding pest resistant rice cultivars were introduced into production agriculture. The continued growth of such cultivars, now in the second or third generation of improvement, has aided several countries in southern and Southeast Asia in meeting the food production needs of their populations.

In spite of its importance in insect pest management programs, plant resistance has not undergone the rapid acceptance and development that has occurred with cultivars possessing disease-resistant qualities. In the past, this was due to the relative ease with which insect control was accomplished by the use of chemical insecticides. Brader (1973) noted that in 1965, 65% of the 300 crop cultivars registered in the United States contained disease resistance, but only 6% contained any significant levels of resistance to insects. The reason for this difference was pointed out by Sprague and Dahms (1972), who indicated that about four times as much pesticide was used for insect control as for disease control during the 1968 growing season. An additional reason for the lag in development of insect resistance is that plant pathogens are more easily cultured for study and inoculation of experimental plant material than are many insect pests. In comparison, insect rearing programs are generally more expensive, may require several years to develop, and may not always produce the behavioral or metabolic equivalent of an insect population in nature.

Since 1978, over 400 cultivars, plant material lines, or parent lines of food and fiber crops have been developed and registered in the United States by the cooperative efforts of entomologists and plant breeders employed by the U.S. Department of Agriculture, state agricultural experiment stations, and private industry (Table 1.1). Well over 100 insect-resistant crop cultivars are grown in the United States today, and probably twice that many insect-resistant cultivars are cultivated in the other major crop production areas of the world. Over one-half of the cultivars developed are those of the major cereal grain food crops of maize, sorghum, and wheat (Fig. 1.1). Many of these cultivars were developed by scientists at international agricultural research centers such as the International Rice Research Institute (IRRI). At IRRI, over 100 rice cultivars have been developed with resistance to all of the major insect pests of rice in southern and Southeast Asia (Table 1.2). Cooperative research efforts between plant breeders and entomologists at IRRI and other international centers have led to a detailed understanding of the type and genetic nature of rice resistance to insects.

There are several reasons why so many more cultivars of insect-resistant cereal and grain crops have been developed than insect-resistant vegetable or fruit crops. The value per acre of the vegetable and fruit crops is usually much lower than that of cereal or grain crops produced in large-scale row crop production systems. Social factors are also responsible for the occurrence of larger numbers of insect-resistant cereal and grain crops. From a social

TABLE 1.1 Development of Germplasm Resistant to Insects in the United States, 1973–1987[a]

Crop	Pest Insect(s)	Germplasm Released		
		Cultivars	Germplasm Lines	Parent Lines
Alfalfa	Alfalfa weevil, spotted alfalfa aphid, blue alfalfa aphid, pea aphid, potato leafhopper, tarnished plant bug	40	28	0
Apple	Woolly apple aphid	1	0	0
Asparagus	Asparagus aphid	1	0	0
Barley	English grain aphid, greenbug, *Rhopalosiphum padi*	4	0	0
Cabbage	Cabbage looper, diamondback moth, imported cabbageworm	0	0	3
Cotton	Banded-wing whitefly, boll weevil, bollworm, cabbage looper, cotton fleahopper, cotton leaf perforator, pink bollworm, tarnished plant bug	6	46	0
Cowpea	Cowpea curculio	2	3	0
Elm tree	Elm leaf beetle	1	0	0
Lettuce	Green peach aphid	1	0	0
Maize	Corn earworm, corn leaf aphid, european corn borer, fall armyworm, northern corn rootworm, southwestern corn borer, western corn rootworm	2	80	8
Muskmelon	Banded cucumber beetle, spotted cucumber beetle, striped cucumber beetle	1	0	8
Peanut	Potato leafhopper, southern corn rootworm, tobacco thrips, velvet bean caterpillar	1	5	0

TABLE 1.1 (*Continued*)

Crop	Pest Insect(s)	Germplasm Released		
		Cultivars	Germplasm Lines	Parent Lines
Pearl millet	Chinch bug	0	1	0
Raspberry	Raspberry aphid, raspberry fruitworm, small-bodied raspberry aphid	1	0	0
Ryegrass	Argentine stem weevil, billbug, sod webworm	2	0	0
Red clover	Pea aphid, yellow clover aphid	1	0	0
Sorghum	Greenbug, sorghum midge, sugarcane borer	0	127	7
Soybean	Corn earworm, Mexican bean beetle, potato leafhopper, soybean looper, velvet bean caterpillar	2	1	0
St. Augustine grass	Southern chinch bug	1	0	0
Sugarcane	Sugarcane borer	4	0	0
Sunflower	Sunflower moth	0	5	1
Sweet potato	Banded cucumber beetle, elongate flea beetle, *Systena frontalis*, pale striped flea beetle, spotted cucumber beetle, sweet potato flea beetle, sweet potato weevil, sweet potato wireworm, white grubs *Plectris aliena* and *Phyllophaga ephilida*	4	0	8
Tobacco	Tobacco hornworm	0	1	0
Triticale	Hessian fly	1	0	0
Turnips	Turnip aphid	2	0	0
Wheat	Cereal leaf beetle, Hessian fly, wheat stem sawfly	26	28	0

[a] From *Crop Science* and *Hort Science*.

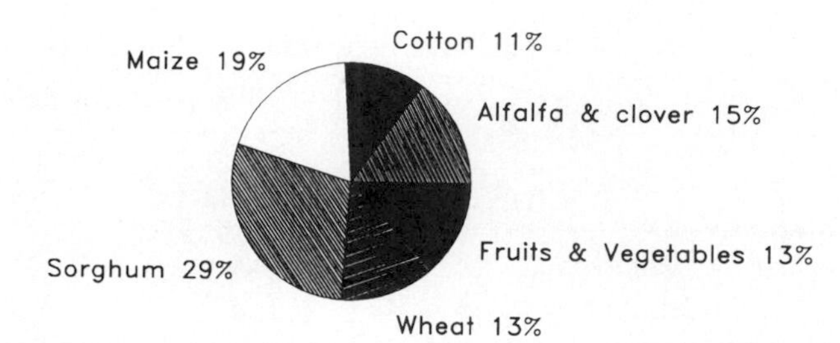

FIGURE 1.1 Breakdown by crop of total insect-resistant germplasm developed in the United States, 1973–1987.

TABLE 1.2 Resistance Category and Number of Genes Involved in Cultivars Identified or Developed to Control the Major Rice Insect Pests of Southeast Asia[a]

Insect	Number of Cultivars	Resistance Genes	
		Category[b]	Number
Striped rice borer	28	Ab	Several
Yellow rice borer	28	Ab	1
Green rice leafhopper	35	Ab, Ax	7
Brown plant hopper	28	Ab, Ax, T	6
White-backed plant hopper	10	Ab; Ax	2
Rice gall midge	22	Ab, Ax	?
Rice whorl maggot	1	Ab, Ax, T	?

[a]From Pathak and Saxena (1980) and Chelliah and Heinrichs (1980).
[b]Ab = antibiosis, Ax = antixenosis, T = tolerance.

perspective, consumers are much more particular about the aesthetic qualities of fruits and vegetables, because they are often eaten or prepared directly, without commercial processing. Thus the consumer is likely to accept an insect-resistant apple cultivar that may harbor fewer insects than susceptibe cultivars, for consumers normally prefer totally insect-free apples. Finally, the economics of research funding also plays a part in the lack of development of insect-resistant fruits and vegetables. Funding for fruit and vegetable insect resistance research has traditionally been lower than that for research on cereal and grain crop insect resistance.

1.2 ADVANTAGES AND DISADVANTAGES OF PLANT RESISTANCE

From an economic vantage point, the use of insect-resistant cultivars in pest management programs offers the grower the advantage of genetically incorporated insect control for the cost of the seed alone. However, if resistance is combined with the use of insecticide, the costs of insecticidal control and insecticide residue problems are generally reduced. Teetes et al. (1986) demonstrated that sorghum hybrids resistant to the sorghum midge *Contarinia sorghicola* (Coquillett) respond more efficiently to insecticide treatment than susceptible hybrids, in terms of net crop value (Fig. 1.2). In addition, insect resistance has also proven to yield much higher returns per dollar invested than those spent on insecticide development. The chemical industry itself predicts a greater use of insect-resistant cultivars in crop protection toward the end of the twentieth century (Table 1.3). Insecticide use is predicted to decline during the same period (Headley 1979). Schalk and Ratcliffe (1976) estimated that approximately 319,000 tons of insecticide (approximately 37% of the total

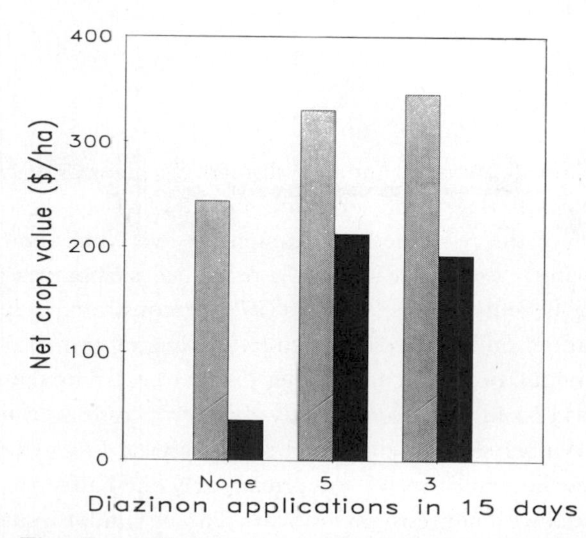

FIGURE 1.2 The effect of diazinon insecticide applications on the net crop value of sorghum hybrids resistant (black bars) or susceptible (gray bars) to the sorghum midge, *Contarinia sorghicola* (Coquillett). Insecticide applied at a rate of 0.6 kg AI/ha five times at 3 day intervals or three times at 5 day intervals. (From Teetes et al. 1986. Reprinted with permission from J. Econ. Entomal. 79: 1091–1095. Copyright 1986, Entomological Society of America.)

TABLE 1.3 Use Trends of Various Pest Control Techniques[a]

Technique	Probable Use in 1982	Trend in Use
CHEMICAL		
Insecticides	Major	Declining
Herbicides	Major	Increasing
BIOLOGICAL		
Predators and Parasites	Minor	No change
Bacteria	Minor	Increasing
Viruses	Not significant	Increasing
Pheromones	Not significant	No change
Plant resistance	Major	Increasing
Insect genetics	Minor	Declining
CULTURAL		
Crop rotation	Minor	Declining
Trap crops	Minor	No change

[a]From Headley (1979).

insecticides applied) are saved annually through the planting of insect-resistant cultivars of corn, barley, grain sorghum, and alfalfa.

The effects of the resistance are cumulative over time, both within and between growing seasons. The longer the resistance is employed and effective, the greater the benefits of its use. Panda (1979) demonstrated reductions in pest insect populations on insect-resistant cultivars ranging from 1.5- to 152-fold. The average population reduction among 25 pest insects damaging 10 food and fiber crops was 12-fold. In a 10-year study of rice crop losses to insect pests in the Philippines, Waibel (1987) determined that rice yield losses of crops planted with insect-resistant cultivars were approximately one-half (14%) of the losses in crops planted with nonresistant cultivars (26%). Similar results were noted during a 4-year study in Java, where insect-related rice yield losses averaged 49% on a susceptible cultivar, but ranged from 27–32% on several resistant cultivars (IRRI 1984).

Presently, plant resistance to insects is used as an integral part of insect pest management programs in rice and many other crops in the world. Plant resistance may be used to enhance chemical control, resulting in reduced rates of insecticide application and ultimately less chemical placed in the environment. For example, morphological changes in improved cotton cultivars that adversely affect insects, such as the split or okra leaf character and the twisted or frego bract character, are currently in use to improve the efficiency of insecticide penetration into the cotton canopy.

In many instances, resistant cultivars synergize the effects of insect biological control agents that suppress pest insect populations. By decreasing the vigor and physiological state of the pest insect, resistant cultivars improve the search efficiency of predators and parasites and enhance the effectiveness of insect pathogens. For example, cultivars of rice resistant to the brown plant hopper, *Nilaparvata lugens* Stl, a major rice pest in Asia, double the efficiency of hopper predators.

In some cases, however, insect-resistant cultivars are not always compatible with biological control agents. In insect-resistant cultivars of several crops, plant trichomes have detrimental effects on the biology of beneficial insect predators and parasites. High levels of insect resistance in some soybean cultivars also have detrimental effects on several species of parasitic Hymenoptera that lower the populations of soybean pest Coleoptera and Lepidoptera species (see Section 10.4.3).

Cultural practices useful in insect pest management also work synergistically with insect-resistant cultivars. The planting of early maturing, insect-resistant rice and soybean cultivars has been shown to reduce populations of several key pest insects. Trap cropping, a practice used to attract pest insect populations and then destroy them, has also been demonstrated to function synergistically when used in combination with insect-resistant cultivars of cotton, soybean, and rice.

Insect-resistant cultivars may also aid in controlling the spread of plant diseases vectored by insects. If the virus is a persistent virus, resistance to the vector usually causes a reduction in the spread of the virus by slowing its population growth (see reviews by Kennedy 1976 & Maramorosch 1980). Thus insect resistance usually integrates effectively with most commonly utilized pest management tactics.

The development and use of insect-resistant cultivars is not without disadvantages. Development of a cultivar resistant to a single insect species normally requires three to five years, and may require ten years or longer for a complex of several insects. Many crops are grown over broad geographic ranges, encompassing widely diverse soil types and environmental conditions. Therefore, different resistant cultivars may be required for different geographic regions. From the plant breeder's perspective, regional crop resistance to insects may be an expensive and time-consuming objective.

Resistance is also commonly identified in wild, undomesticated species of plants that may have only distant taxonomic relations to the crop species under improvement. It is not unusual for these plants to have poor yield, poor plant type, or disease susceptibility. With the adaptation and use of genetic enegineering techniques in crop plants, some of these problems may be eliminated (see Chapter 6). For the most part, however, the incorporation of insect resistance from wild species of plants into domestic crop plants is a long-term process.

Finally, insect-resistant cultivars that rely on the effects of a single, major gene often promote the development of resistance-breaking biotypes. The use of monogenic resistance usually leads to a pattern of sequential cultivar (gene) release, with each new cultivar having a different gene or gene arrangement, in order to stay a step ahead of the continuously mutating genetic machinery of the pest insect. The development of cultivars with polygenic resistance that delays biotype development requires several years longer to accomplish. Rice cultivars with resistance to stem-boring Lepidoptera have been used in Asian rice production systems for several years without the development of resistance-breaking borer biotypes (Heinrichs 1986). Efforts are currently in progres to develop polygenic resistance in cotton, wheat, and rice to other pest insects.

1.3 DEFINITIONS

By definition, *plant resistance to insects* is composed of the genetically inherited qualities that result in a plant of one cultivar or species being less damaged than is a susceptible plant, which lacks these qualities. Plant resistance to insects is always relative; the degree of resistance is based on comparison to susceptible plants that are more severely damaged under similar test conditions. This practice is necessary, since resistance is influenced by environmental fluctuations occurring over both time and space. In the terms of the plant resistance researcher, susceptibility is the inability of a plant to inherit qualities that convey resistance.

Pseudo-resistance or *false resistance* may also occur in normally susceptible plants for several different reasons: Plants may avoid insect attack because they were planted earlier than normally. Resistance may also be induced temporarily by variations in temperature, daylength, soil chemistry, plant or soil water content, or internal plant metabolism. Finally, susceptible plants may simply escape damage because of incomplete insect infestation.

Associational resistance occurs when a normally susceptible plant grows in association with a resistant plant and derives protection from insect predation from the resistant plant. The population density of *Diaphania hyalinata* (L.), a lepidopteran pest of cucurbit crops in Mexico, is much lower in squash polyculture (squash–maize–cowpea) than in squash monoculture (Letourneau 1986). High fruiting strains of cotton, planted in combination with cultivars not preferred for oviposition, have been used to attract and hold boll weevil infestations away from commercial plantings of cotton (Burris et al. 1983). Associational resistance indicates that the diversionary or delaying actions of mixtures of plant species can help slow the development of insect biotypes (Chapter 9) that overcome insect-resistant cultivars. From a practical standpoint, associational resistance may be implemented by the development of insect-resistant cultivars based on several different sources of resistance. For expanded discussions of this topic, see Chapters 8 and 10.

Three modalities of plant resistance to insects, termed *functional categories* by Horber (1980), are commonly referred to in plant resistance literature. These terms were originally defined by Painter (1951). The effects of resistant plants on insects can be manifested as *antibiosis*, in which the biology of the pest insect is adversely affected, or as *antixenosis*, in which the plant acts as a poor host and the pest insect then selects an alternate host plant. The inherent genetic qualities of the plant itself may aid it in expressing *tolerance* to the pest insect and afford it the ability to withstand or recover from insect damage. These terms have been accepted because of conceptual convenience, but they are not always biologically discrete entities.

Often there is overlap between the antibiosis and antixenosis categories of resistance because of the difficulty involved in designing experiments to delineate between the two. For example, if an insect confined to a resistant plant fails to gain weight at the rate it normally does on a susceptible plant, it might be assumed that lack of weight gain in due to the presence of antibiotic properties in the plant. However, the lack of weight gain may also be due to the presence of a physical or chemical feeding deterrent with strong antixenotic properties. This deterrent may initiate aberrant behavior in the test insect, which results in a weakened physiological condition. Quite often, very detailed sets of experiments are required to delineate the actual contributions of plant factors to each category of resistance. It is not unusual to find that combinations of each category are responsible for insect resistance. From a practical standpoint, the absolute contribution of a given category may never be fully elucidated. Detailed descriptions and discussions of each of these categories and the methodologies involved in investigating them are presented in Chapters 2, 3, 4, and 6.

In summary, the practice of identifying and cultivating plants with insect-resistant qualities is an ancient one that has been increasingly accepted in modern crop pest management systems. This acceptance has been aided by the continual development of insect populations with genetic resistance to chemical pesticides and by a continual need to produce crops that require progressively lower production costs. As was previously indicated, a major advantage in the cultivation of insect-resistant crops is that production costs are lower, because some or all of the insect control costs are incorporated into the seeds or clones themselves. In the following chapters we investigate how this control is identified, the techniques used to manipulate it, and how it can ultimately be used to manage arthropod populations in crop pest management systems.

REFERENCES

Beck, S. D. 1965. Resistance of plants to insects. Ann. Rev. Entomol. 10:107–232.

Brader, L. 1973. Ecological basis for insect control—Host plant resistance. Proceedings, FAO Conference on Ecology in Relation to Plant Pest Control, *Food Agriculture Organization of the United Nations. Rome, Italy*, pp. 55–65.

Burris, E., D. F. Clower, J. E. Jones, and S. L. Anthony. 1983. Controlling boll weevils with trap cropping, resistant cotton. La. Agric. 26:22–24.

Chapman, I. 1826. Some observations on the Hessian fly; written in the year 1797. Mem. Phil. Soc. Prom. Agr. 5:143–153.

Chelliah, S. and E. A. Heinrichs. 1980. Varietal resistance of rice to insect pests. Unit PC-10, Rice Production Training Series. International Rice Research Institute, Los Banos, Philippines, 24 pp.

Dicke, F. F. 1972. Philosophy on the biological control of insect pests. J. Environ. Qual. 1:249–254.

Green, M. B. and P. A. Hedin. 1986. *Natural Resistance of Plants to Pests*. ACS Symposium Series 296. American Chemical Society, Washington, DC, 243 pp.

Harris, M. O. 1980. Biology and Breeding for Resistance to Arthropods and Pathogens in Agricultural Plants. Texas Agric. Exp. Sta. Publ. MP-1451.

Havens, J. N. 1792. Observations on the Hessian fly. Trans. N.Y. Soc. Agron. Pt. 1:89–107.

Headley, J. C. 1979. Economics of pest control: Have priorities changed? Chem. Eng. News, Jan. 15, 1979, pp. 55–57.

Hedin, P. A. (ed.) 1978. *Plant Resistance to Insects*. ACS Symposium Series 62. American Chemical Society, Washington, DC, 286pp.

Hedin, P. A. (ed.) 1983. *Plant Resistance to Insects*. ACS Symposium Series 208. American Chemical Society, Washington, DC, 374 pp.

Heinrichs, E. A. (1986). Perspectives and directions for the continued development of insect-resistant rice varieties. Agric. Ecosyst. Environ. 18:9–36.

Horber, E. 1980. Types and classification of resistance. In F. G. Maxwell and P. R. Jennings (eds.). *Breeding Plants Resistant to Insects*. Wiley, New York, pp. 15–21.

International Rice Research Institute. 1984. Entomology. In *A Decade of Cooperation and Collaboration Between Sukamandi (AAARD) and IRRI 1972–1982*. International Rice Research Institute, Los Banos, Philippines, 169 pp.

Kennedy, G. G. 1976. Host plant resistance and the spread of plant diseases. Environ. Entomol. 5:827–832.

Letourneau, D. K. 1986. Associational resistance in squash monocultures and polycultures in tropical mexico. Environ. Entomol. 15:285–292.

Lindley, G. 1831. *A guide to the Orchard and Kitchen Garden*. Longman, Rees, Orme, Brown & Green Publ. Co., London, 597 pp.

Maramorosch, K. 1980. Insects and plant pathogens. In F. G. Maxwell and P. R. Jennings (eds.). *Breeding Plants Resistant to Insects*. Wiley, New York, pp. 138–155.

Maxwell, F. G., J. N. Jenkins, and W. L. Parrott. 1972. Resistance of plants to insects. Adv. Agron. 24:187–265.

Maxwell, F. G. and P. R. Jennings (eds.). 1980. *Breeding Plants Resistant to Insects*. Wiley, New York, 683 pp.

Painter, R. H. 1951. *Insect Resistance in Crop Plants*. University of Kansas Press, Lawrence, KS, 520 pp.

Painter, R. H. 1958. Resistance of plants to insects. Ann. Rev. Entomol. 3:267–290.

Panda, N. 1979. *Principles of Host-Plant Resistance to Insect Pests*. Allanheld, Osmun and Universe Books, New York, 386 pp.

Pathak, M. D. and R. C. Saxena. 1980. Breeding approaches in rice. In F. G. Maxwell and P. R. Jennings (eds.). *Breeding Plants Resistant to Insects*. Wiley, New York, pp. 421–455.

Russell, G. E. 1978. *Plant Breeding for Pest and Disease Resistance*. Butterworth, Boston, 485 pp.

Schalk, J. M. & R. H. Ratcliffe. 1976. Evaluation of ARS program on alternative methods of insect control: Host plant resistance to insects. Bull. Entomol. Soc. Am. 22:7–10.

Snelling, R. O. 1941. Resistance of plants to insect attack. Bot. Rev. 7:543–586.

Sprague, G. F. & R. G. Dahms. 1972. Development of crop resistance to insects. J. Environ. Qual. 1:28–33.

Teetes, G. L., M. I. Becerra, and G. C. Peterson. 1986. Sorghum midge (Diptera: Cecidomyiidae) management with resistant sorghum and insecticide. J. Econ. Entomol. 79:1091–1095.

Waibel, H. 1987. *The Economics of Integrated Pest Control in Irrigated Rice.* A Case Study from the Philippines. Springer-Verlag, Berlin, 196 pp.

2 Antixenosis—The Effect of Plant Resistance on Insect Behavior

2.1 DEFINITIONS AND CAUSES

Antixenosis, a term derived from the Greek word *xeno* (guest), describes the inability of a plant to serve as a host to an insect herbivore. As a result, a potential pest insect is forced to select an alternate host plant. The term *antixenosis resistance* in plants was proposed by Kogan and Ortman (1978) to describe more accurately the *nonpreference* reaction of insects to a resistant plant, which was originally defined by Painter (1951). *Nonpreference resistance* is the group of plant characters that lead to a plant being less damaged than another plant lacking these characters and the insect responses to them.

Both antixenosis and nonpreference denote the presence of morphological or chemical plant factors that adversely alter insect behavior, resulting in selection of an alternate host plant. Physical barriers such as thickened plant epidermal layers, waxy coatings on leaves and stems, or trichomes (plant hairs) may force insects to abandon their efforts to feed on an otherwise palatable host plant. Insect-resistant crop plants may be devoid of, or lack suffient levels of, phytochemicals that stimulate feeding or oviposition, labeling them as "also rans" in the supermarket of insect foods. Antixenosis in plants may also be due to the possession of unique phytochemicals that repel or deter insect herbivores from feeding or oviposition. For a discussion of the types of techniques used to determine the existence of and to quantify antixenosis, see Section 6.2.1.

2.2 INSECT SENSORY SYSTEMS INVOLVED IN HOST SELECTION

To understand how antixenosis functions in a plant's resistance to insects, we can pose the question "How does an insect know where 'home' is?" The answer is determined by an insect's ability to perceive and integrate external stimuli detected by olfactory, visual, tactile, and gustatory receptors. The following

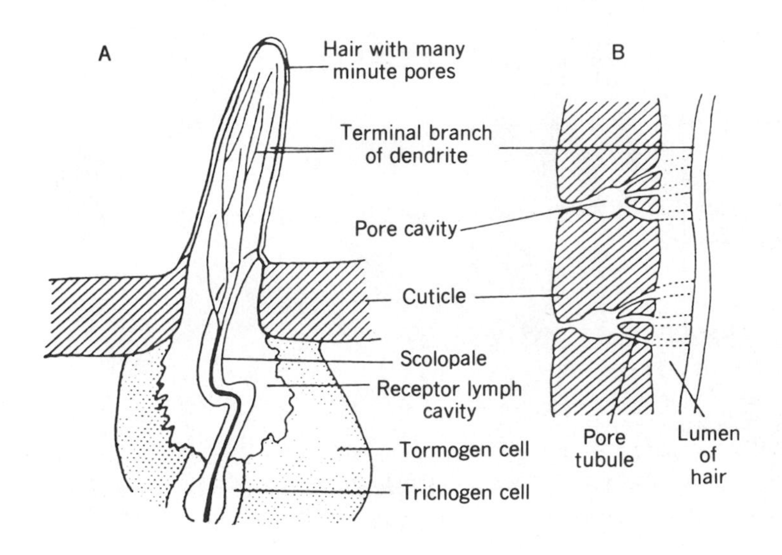

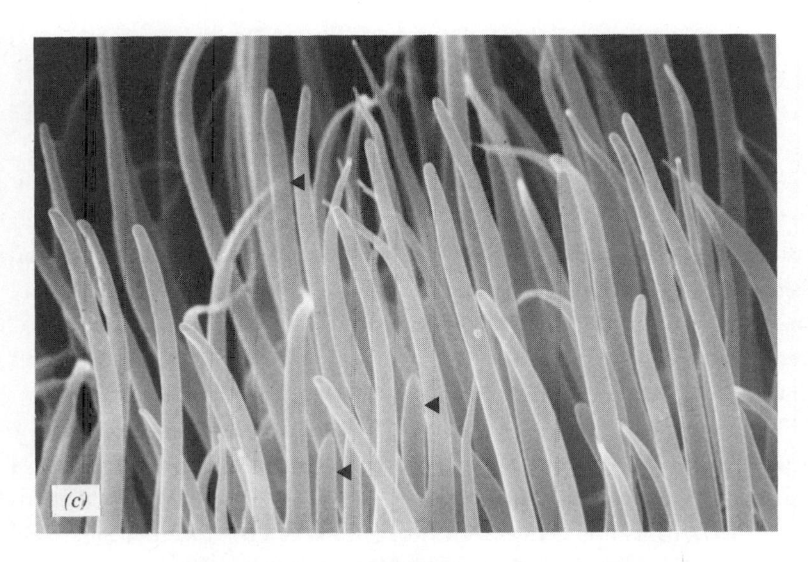

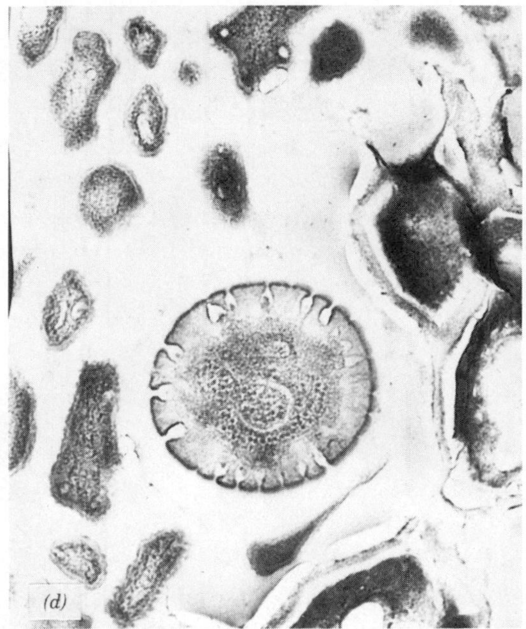

FIGURE 2.1 Insect sensilla basiconica. (A) Diagram of sensillum, (B) exploded diagram of hair wall showing pores through which stimulating molecules reach the nerve (dendrite), both from Chapman and Blaney 1979; (C) sensilla basiconica on the antennal club of the rice water weevil, *Lissorhoptrus oryzophilus* Kuschel, (4088 ×); (D) cross section of sensillum basiconicum on the antennal club of *L. oryzophilus* showing pores in sensillum wall (35,200 ×). (Figure 2.1*a* and *b* reproduced with the permission of Academic Press, Inc.)

sections describe the rudimentary basis of each type of stimuli. More detailed discussions are presented in a recent volume by Ahmad (1983).

2.2.1 Olfaction

In order to perceive the plant odors emitted by potential host plants, insects rely on an olfactory guidance system controlled by cuticular sense organs known as sensilla basiconica, located on the antennae. Sensilla basiconica are porous, thin-walled structures that range in length from 10 to 20 μm. (Fig. 2.1). Great diversity exists in the number and arrangement of these sensible on the antennae of various insects.

The olfactory sensitivity of different insect species is instinctively tuned to and controlled by a given qualitative and quantitative blend of odors. Most plant species are unique in their composition of volatile phytochemicals produced by roots and leaves. Specific groups of odor components in plants such as the sulfur compounds in onion (Matsumoto 1970, Pierce et al. 1978) and Leeks (Lecomte & Thibout 1981), the propenylbenzenes in carrots (Guerin et al. 1983), and the leaf alcohols in potato plants (Visser et al. 1979) play an important role in directing insect movement to a host plant. In some cases, specific olfactory sensilla are known to respond to specific odor components of a plant's odor "bouquet" (Mustaparta 1975).

Dethier et al. (1960) defined the effects of phytochemicals based on the responses elicited by insects. Odors emitted by plants that stimulate insect olfactory receptors and cause long-range insect movement toward odors are termed *attractants*. Plants exhibiting antixenosis may produce olfactory *repellents*, which cause insects to move away from the plants producing the odor. Susceptible plants may also emit *arrestment odors*, which cause insect movement to cease in close proximity to the odor source. The interplay between the odors emitted by plant sources, the effects of the environment on these odors, the perception of the odors by insects, and the resultant insect behaviors have been summarized by Visser (1986) (Fig. 2.2). Thiery and Visser (1987) have demonstrated how the attraction of odors of potato foliage to the Colorado potato beetle, *Leptinotarsa decemlineata* (Say), are neutralized in an odor blend of potato and the wild nonhost tomato, *Lycopersicon hisutum* f. *glabratum*.

2.2.2 Vision

Prokopy and Owens (1983) describe vision in herbivorous insects as functioning in two ways. The first involves the ability of insects of perceive spatial patterns

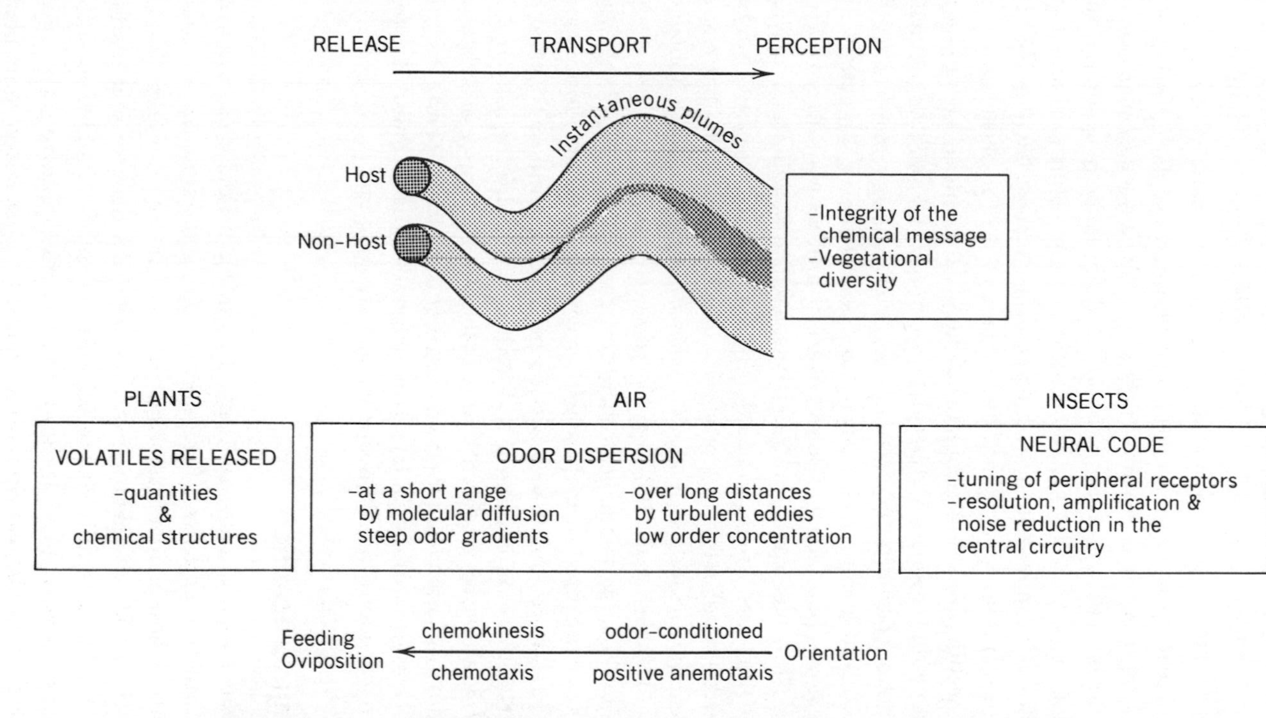

FIGURE 2.2 Schematic representation of the release of plant volatiles, their dispersion and perception by insects, and resultant insect behaviors. (From Visser 1983. Reproduced with permission from the *Annual Review of Entomology*, Vol. 31. Copyright 1986 by Annual Reviews Inc.)

using instinctive stimuli "templates." The second process involves an insect's ability to detect differences in the brightness, hue, and saturation of various wavelengths of light. Visual cues received by insects during host plant searching and location result from their perception of the *spectral quality* of the light stimuli, the *dimensions* of the objects viewed, and the pattern or shape of the object.

Visual and chemical stimuli are perceived simultaneously during the orientation of an insect to a potential host plant. During long-range orientation, an insect may use vision for recognition of the shape of an object and utilize olfaction to perceive plant attractants. After approaching the immediate location of the plant, movement to the plant surface is most likely guided by perception of the outline of the plant. Final contact with the plant surface by many foliage feeding insects (Coleoptera, Diptera, Homoptera, Lepidoptera, Thysanoptera) is due to a positive response to yellow or yellow-green pigments in plant foliage that occur in the spectral range of 500–580 nm. (see Fig. 6.19).

Antixenois resistance in crop cultivars has been achieved by genetically altering the color of plant foliage. Some cucurbit cultivars with silver colored leaves reflect more blue and ultraviolet wavelengths of light than normal cultivars, and are resistant to aphids and aphid vectored diseases (Shifriss 1981). The red leaf color trait in cotton is a heritable character that causes antixenotic reactions in adult boll weevils, *Anthonomous grandis grandis* Boheman (Iseley 1928). Reinhert et al. (1983) evaluated over 40 cultivars of *Canna* spp. for resistance to the larger canna leafroller, *Calpodes ethlius* (Stoll). Plants with red leaf color are preferred over those with green leaves for oviposition, and cultivars with yellow or pink flowers are less preferred than those with red or orange flowers for oviposition and larval feeding. The color intensity of leaf supernatants of birch trees has been used to determine the degree of resistance to oviposition by the birch leafminer, *Fenusa pusila* (Lepeletier) (Fiori & Craig 1987). Birch species with high levels of oviposition have lower spectrophotometric absorption rates than species that are resistant.

2.2.3 Thigmoreception

After an insect makes contact with a plant surface, trichoid sensilla on the body, tarsi, head, and antennae (Fig. 2.3) perceive tactile stimuli and supply information about host plant morphology to the insect nervous system. Stimuli are received from leaf or stem trichomes, epidermal ridges, or leaf margin notches that trigger genetically controlled sequences of feeding or oviposition behavior. Plant morphological features may promote positive mechanical stimuli and act as feeding or oviposition stimulants. Conversely, some plant

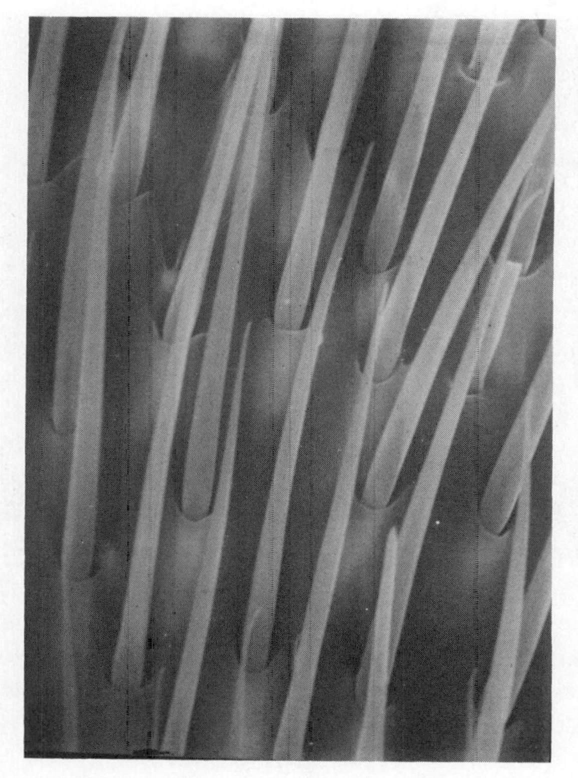

FIGURE 2.3 Sensilla trichoidea on the antennal club of the clover head weevil, *Hypera meles* (F.) (2000 ×). (From Smith et al. 1976. Reprinted with permission from *Int. J. Insect Morphol. Embryol.* 5. Copyright 1976, Pergamon Press, Inc.)

morphological features may prevent or disrupt the normal mechanoreception process, and thus function as deterrents of feeding and oviposition deterrents.

2.2.4 Gustation

Insects taste plant tissues with contact chemosensory sensilla styloconica, maxillary palpi (Fig. 2.4), and labral gustatory receptors that also transmit sensory information to the insect central nervous system. At the tip of these sensilla, a single pore, open to the environment, receives chemosensory stimuli from molecules of plant allelochemicals in the liquid or vapor phase (Fig. 2.5). Stimuli provided by these molecules are then electrochemically transduced and transmitted to the insect central nervous system (Hanson 1983). Several

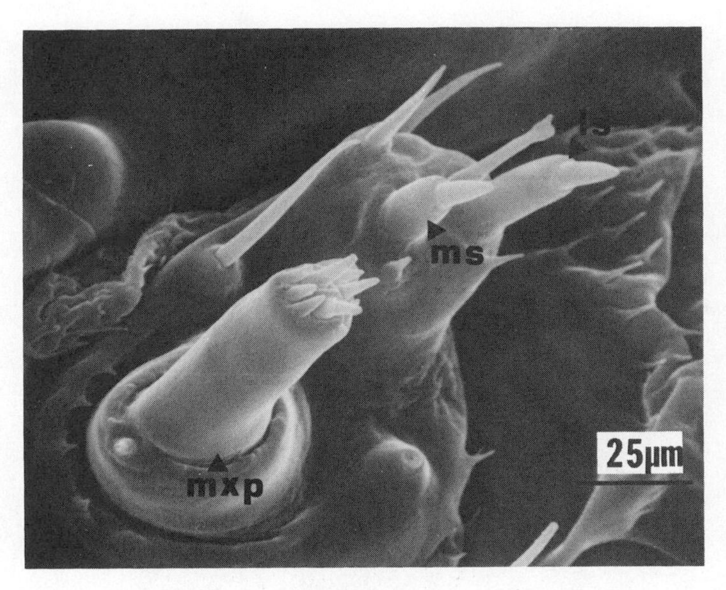

FIGURE 2.4 Maxilla of last stage tobacco budworm, *Heliothis virescens* (F.), larvae. ls = lateral sensillum styloconicum; ms = medial sensillum styloconicum; mxp = maxillary palp. (From Baker et al. 1986. Reprinted with permission from *Int. J. Insect Morphol. Embryol.*, 15. Copyright 1986, Pergamon Press, Inc.)

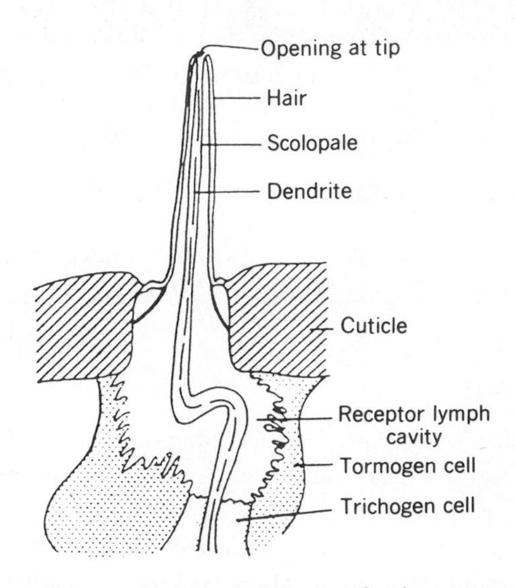

FIGURE 2.5 Longitudinal section through the tip of an insect contact chemoreceptor. (From Chapman and Blaney 1979. Reproduced with the permission of Academic Press, Inc.)

different types of gustatory receptors may determine both qualitative and quantitative differences in the chemical content of the plant tissues tasted. The response spectrum of the gustatory receptor of an insect depend on the distribution of the different types of phytochemicals within those plants of an insect's host range. The receptors of generalist (oligophagous) insects normally have a wider response spectrum than the receptors of specialist (monophagous) insects (Visser 1983). In a review of the sensory response of the mouthpart sensilla of grasshoppers, Chapman (1988) concludes that not all sensilla need be stimulated for distinction of acceptable versus unacceptable food and that monophagous grasshoppers require fewer sensilla to select food than polyphagous species. Phytochemical stimuli that elicit continued insect feeding after being tasted are known as *phago* (feeding) *stimulants*, and those stimuli that prevent feeding are referred to as *phagodeterrents*. Although less commonly used, the terms feeding *incitants* and *suppressants* are used to denote stimuli that initate or prevent feeding, respectively (Dethier et al. 1960). A representative description of the outcome of the perception of all plant stimuli and the resulting insect responses (Visser 1983) (Fig. 2.6) indicates that the total sensory input from chemical, visual, and tactile plant stimuli determine the ultimate degree of acceptance and utilization of a plant by an insect.

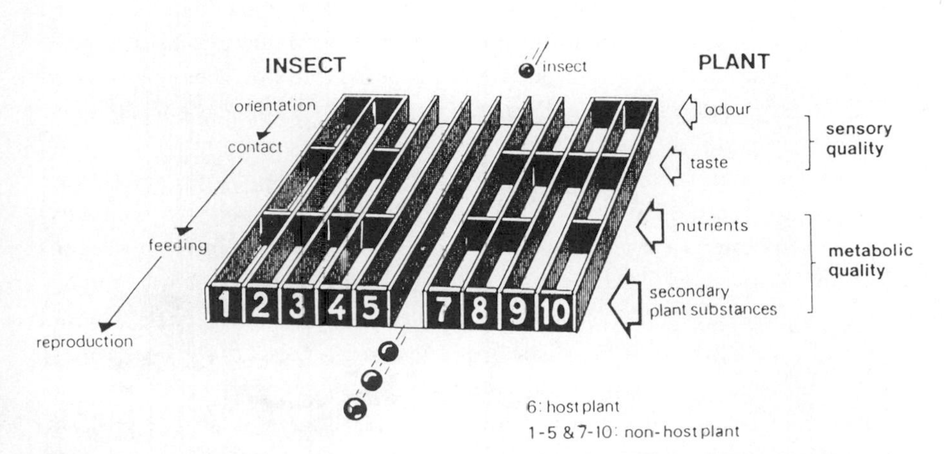

FIGURE 2.6 "Marble box" perception of plant and environmental stimuli by insects and resulting host or nonhost selection (From Visser 1983. Reprinted with permission from P. A. Hedin (ed.). *Plant Resistance to Insects*. Copyright 1983, American Chemical Society.)

2.3 HISTORY OF THEORIES OF HOST SELECTION BY INSECTS

Humans have observed the behavior and life history of insects for centuries. Not surprisingly, several theories have been advanced since the beginning of the twentieth century about how an insect locates a plant. The first to offer a postulate on insect host selection was Brues (1920), who proposed a *botanical instinct theory*, which suggested that insects select host plants that meet specific nutritional and ecological requirements of an insect not offered by other plant species. Brues viewed the acceptance of a particular plant by an insect to be related to the nutritional composition and ecological niche of the plant in an insect's environment. The *token stimuli theory*, proposed by Fraenkel (1959), reasoned that insect host plant selection is determined by specific *secondary plant substances* or phytochemicals, such as glycosides, phenols, tannis, alkaloids, terpenoids, and saponins. Fraenkel suggested that these chemicals are nonnutritional and that their sole purpose is to defend plants against phytophagous insects and diseases. He theorized that insects evolve from polyphagy to monophagy to overcome the adverse effects of secondary plant substances and that they begin using these compounds as cues to locate acceptable host plants.

Since then, studies by Seigler and Price (1976) and Heftmann (1975) have demonstrated a high turnover rate of secondary plant substances in plants and their close association with primary metabolic functions. These substances also function as regulators of important biochemical plant processes. Therefore, their functions as plant metabolites appear to be more primary than secondary.

Prior to the discoveries about the metabolism of secondary plant substances, Kennedy (1965) and Thorsteinson (1960) proposed that host selection is based on insect responses to both nonnutrient and nutrient phytochemicals. Their theories stemmed from the fact that many insects can be stimulated to feed by nutrient chemicals such as amino acids, carbohydrates, and vitamins (House 1969, Hsiao 1969). The *dual discrimination theory*, proposed by Kennedy, continues to dominate contemporary views on the subject of insect host plant selection. Whittaker (1970) introduced the term *allelochemical* to replace secondary plant substances. He defined an allelochemical as a "non-nutritional chemical produced by an individual of one species that affects the growth, health, behavior, or population biology of another species." Scores of allelochemicals have been shown to affect insect feeding behavior, oviposition, and survival in either a positive or negative manner.

Allelochemicals can function as *allomones*, of benefit to the producing organism, in this case a resistant plant. They also act as *kairomones*, where they benefit the insect recipient. In terms of plant resistance to insects, allomones represent plant-produced deterrents, repellents, or inhibitors of feeding and oviposition. Conversely, the attractants, arrestants, and feeding or oviposition stimulants found in susceptible plants function as kairomones.

2.4 EFFECTS OF PLANT DEFENSES ON INSECT BEHAVIOR

2.4.1 Plant Morphology

Several types of plant morphological defenses are employed by plants to deter insect feeding and oviposition. Surface hairs or trichomes are the first plant organs contacted during the preliminary stages of host acceptance. Levin (1973) and Webster (1975) comprehensively reviewed the effects of plant pubescence on insects. In addition, several more recent examples of the effects of plant pubsencence on insects (Table 2.1) are presented in the following discussion.

2.4.1.1 Trichomes

Dense growths of simple erect trichomes on leaves of some soybean (Fig. 2.7*a*) and alfalfa cultivars deter feeding by the potato leafhopper, *Empoasca fabae* Harris (Taylor 1956, Lee 1983). Trichome-based antixenosis also exists in cotton cultivars resistant to the leafhopper, *Empoasca facialis* Jacobi (Reed 1974). These trichomes limit the access of hopper feeding stylets to plant tissues and interfere with hopper attachment to the plant. Glandular haired alfalfa species also exhibit antixenosis to the potato leafhopper (Shade et al. 1979), the alfalfa weevil, *Hypera postica* Gyllenhal (Johnson et al. 1980a, b, Danielson et al. 1987), the alfalfa seed chalcid, *Bruchophagus roddi* (Gussakovsky) (Brewer et al. 1986), the spotted alfalfa aphid, *Therioaphis maculata* (Buckton) (Ferguson et al. 1982), the alfalfa blotch leaf miner, *Agromyza frontella* (Rondani) (MacLean & Byers 1983), and the pea aphid, *Acrythrosiphon pisum* (Harris) (Shade & Kitch 1983).

Glandular trichomes (Fig. 2.7*b*) on the wild potato, *Solanum neocardenasii*, adversely affect the feeding behavior of the green peach aphid, *Myzus persicae* (Sulzer), by delaying the amount of time required to begin feeding (Lapointe & Tingey 1986). (see Section 3.2.2.2 for a discussion of the different trichome types.) Tingey and Laubengayer (1986) demonstrated that removal of pubescence increases feeding by the potato leafhopper. Similarly, leaf pubescence also contributes to the feeding antixenosis of some cultivars of soybean to

TABLE 2.1 Insects Affected by Resistant Crop Plant Cultivars Possessing Leaf and Stem Trichomes

Crop Plant	Trichome Type	Insect Affected	Reference
Alfalfa	Simple	Potato leafhopper	Shade et al. 1979
	Glandular	Alfalfa weevil	Danielson et al. 1987
		Alfalfa seed chalcid	Brewer et al. 1983
		Spotted alfalfa aphid	Ferguson et al. 1982
		Pea aphid	Shade and Kitch 1983
		Alfalfa blotch leaf miner	MacLean and Byers 1983
Cotton	Simple	*Empoasca fasciatus*	Reed 1974
		Boll weevil	Wessling 1958
		Cotton leafworm	Kamel 1965
		Tarnished plant bug	Meredith and Schuster 1979
		Lygus hesperus	Benedict et al. 1983
Poinsettia	Simple	Whitefly	Bilderback and Mattson 1977
Potato	Simple	Potato leafhopper	Taylor 1956
	Glandular	Green peach aphid	Lapointe and Tingey 1986
Sorghum	Simple	Sorghum shoot fly	Maiti and Gibson 1983
Soybean	Simple	Agromyzid bean flies	Chiang and Norris 1983
		Cabbage looper	Khan et al. 1986
		Potato leafhopper	Lee 1983
Strawberry	Simple	Black vine weevil	Doss et al. 1987
Sugarcane	Simple	*Scirpophaga nivella*	Verma and Mathur 1950
Sunflower	Glandular	Sunflower moth	Rogers et al. 1987
Wheat	Simple	Bird cherry oat aphid	Roberts and Foster 1983
		Cereal leaf beetle	Hoxie et al. 1975
		Hessian fly	Roberts et al. 1979

FIGURE 2.7 (*a*) Simple trichomes of soybean leaves which cause antixenosis to cabbage looper, *Trichoplusia ni* Hubner, larvae (6 ×). (From Khan et al. 1986. Reprinted with the permission of D. W. Junk Publishing Co.). (*b*) Glandular trichomes of potato foliage imparting resistance to the green peach aphid, *Myzus persicae* (Sulzer) (120 ×). (From Ryan et al. 1982. Reprinted with permission from *Phytochemistry* 21. Copyright 1982, Pergamon Press, Inc.)

(a)

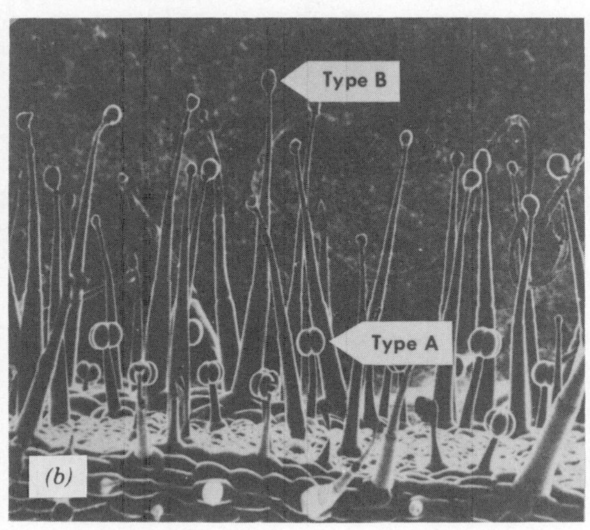
Type B
Type A
(b)

the cabbage looper, *Trichoplusia ni* (Hubner) (Khan et al. 1986), and agromyzid bean flies (Chiang & Norris 1983).

High densities of trichomes on the buds of some cotton cultivars also deter feeding and oviposition by the boll weevil, *Anthomonus grandis grandis* Boheman (Wessling 1958). Kamel (1965) determined that cotton cultivars with increased trichome density on lower leaf surfaces were more resistant to the cotton leafworm, *Spodoptera littoralis* (Boisd.). Pubescent cotton cultivars also exhibit antixenosis to *Lygus lineolaris* Palisot de Beauvois (Meredith & Schuster 1979, Wilson & George 1986), *Lygus hesperus* Knight (Benedict et al. 1983), and the cabbage looper (George et al. 1977). Pubescent cotton promotes population growth of the cotton fleahopper, *Pseudatomoscelis seriatus* (Reuter) (Lukefahr et al. 1970) and tobacco budworm, *Heliothis virescens* (F.) (Lukefahr et al. 1971). For these insects, cotton cultivars with smooth-leaved foliage with little or no pubescence are resistant (Robinson et al. 1980).

In contrast, a glabrous pearl millet isoline has a negative effect on both oviposition and feeding by fall armyworm, *Spodoptera frugiperda* (J. E. Smith), adults and larvae (Burton et al. 1977). Maize selections of the 'Antigua' ancestry, with reduced trichome density and delayed seeding development of pubsence, are less preferred for oviposition by bollworm larvae and posses resistance to feeding (Wiseman et al. 1976, Widstrom et al. 1979).

Certain wheat cultivars and *Avena* (oat) species with dense growth of long, erect trichomes deter oviposition by the cereal leaf beetle, *Oulema melanoplus* (L.), much more than cultivars with sparse growth of short trichomes (Hoxie et al. 1975, Wallace et al. 1974). The pubescent wheat cultivar 'Vel' also exhibits antixenosis to adults and larvae of the Hessian fly, *Mayetiola destructor* (Say) (Roberts et al. 1979). Adult flies lay fewer eggs, egg hatch is reduced, and larval mobility is impaired. Some pubescent wheat cultivars also have antixenotic effects on the bird cherry oat aphid, *Rhopalosiphum padi* (L.), an important vector of barley yellow dwarf virus (Roberts & Foster 1983). However, pubescent wheats are more susceptible than glabrous wheats to the wheat curl mite, *Eriophyes tulipae* Keifer, (Harvey & Martin 1980).

Foliar pubescence is also responsible for antixensions in cultivars of sorghum resistant to the sorghum shootfly, *Atherigona soccata* (Maiti & Gibson 1983), and in poinsettia cultivars that are resistant to the whitefly, *Trialeurodes vaporariorum* (Westwood) (Bilderback & Mattson 1977). Doss et al. (1987) determined that the resistance of strawberry clones to feeding and oviposition by the black vine weevil, *Otiorhynchus sulcatus* (F.), is related to the density of simple trichomes on the underside of leaves. Leaves of the composite plant *Anaphalis margaritacea* contain long, dense trichomes that impart resistance to the meadow spittlebug, *Philaenus spumarius* (L.) (Hoffmann & McEvoy 1985).

2.4.1.2 Surface Waxes

Plant leaves are protected against desiccation, insect predation, and disease by a layer of surface waxes over the epicuticle. Chemically, plant waxes are esters formed by the linkage of a long-chain fatty acid and an aliphatic alcohol. For a more in-depth discussion of the structural aspects of plant surface waxes, see Jeffree (1986).

Foliar wax coatings play an important role in the resistance of some crop cultivars to insect attack (Table 2.2) when sense organs on the insect tarsi and mouthparts receive negative chemical and tactile stimuli from the leaf surface. The raspberry species *Rubus phoenicolasius* has heavy wax secretions that serve as a resistance barrier against the raspberry beetle, *Byturus tomentosus* Barber, and the rubus aphid, *Amphorophora rubi* (Kaltenbach) (Lupton 1967). Wax blooms on the leaves of some cruciferous crops deter feeding of the cabbage flea beetle, *Phyllotreta albionica* (LeConte) (Anstey & Moore 1954). Wax blooms on the leaves of brussels sprouts consist of vertical rods and dendritic plates that interfere with adhesion of the tarsal setae of the mustard beetle, *Phaedon cochleariae*, to the leaf surface (Stork 1980). In contrast, the waxy leaves of kale stimulate feeding of the cabbage aphid, *Brevicoryne brassicae* (L.), and the cabbage whitefly, *Aleurodes brassicae* (Walker) (Thompson 1963) more than glossy-leaved cultivars.

TABLE 2.2 Effects of Plant Surface Waxes on Antixenosis Resistance to Insects

Plant	Insect	Wax Effect(s)	Reference
Brussels sprouts	Flea beetle	Adhesion	Stork 1980
Crucifers	Flea beetle	Deterrent	Anstey and Moore 1957
Kale	Cabbage aphid	Feeding stimulant	Thompson 1963
Raspberry	Raspberry aphid and beetle	Physical barrier	Lupton 1967
Sorghum	Locust	Deterrent	Atkins and Hamilton 1982
	Greenbug aphid	Attractant/ stimulant	Weibel and Starks 1986

Foliar surface waxes of other plants contain allelochemicals that also affect the behavior of insects. Sulfur compounds from the surface waxes of onions stimulate oviposition by the leek moth, *Acrolepiopsis assectella* (Zeller) (Thibout et al. 1982). Linear furanocoumarins from carrot foliage waxes have a similar effect on oviposition by the carrot rust fly, *Psilia rosae* (F.) (Stadler & Buser 1982). Hydrocarbon- and carbonyl-containing fractions of the wax of rice cultivars resistant to the brown planthopper, *Nilaparvata lugens* Stal, adversely alter hopper feeding behavior by causing hoppers to move away from their preferred feeding site (Woodhead & Pradgham 1988).

In sorghum, a parallel exists to that in cruciferous plants. Wax on sorhum leaves deters feeding by the migratory locust, *Locusta migratoiodes* (R & F.) (Atkin & Hamilton 1982). Wax from young plants is more deterrent than that from older leaves, apparently because of to the wax "blooms" of younger, faster growing foliage (Fig. 2.8). In contrast, sorghum isolines without leaf wax blooms exhibit antixenosis to the greenbug, *Schizaphis graminum* Rodani (Peiretti et al. 1980, Weibel & Starks 1986). This quality is not present in seedlings, but develops during the later stages of vegetative plant development.

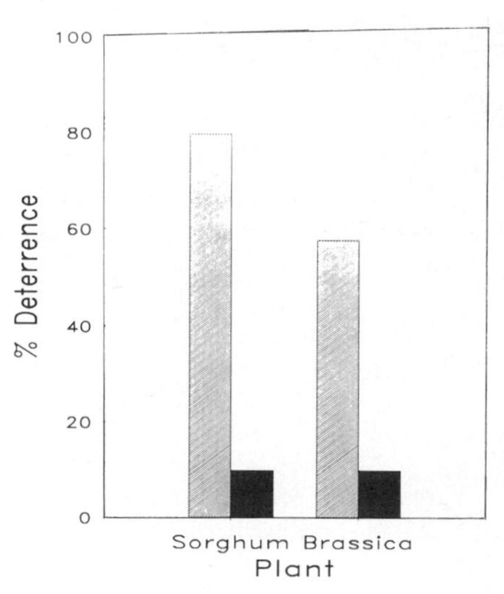

FIGURE 2.8 Deterrence (% insects stopping at palpation) of *Locusta migratoria* nymphs by surface waxes on plant foliage. (Adapted from Woodhead and Chapman 1986. Reprinted with permission of Edward Arnold Publishing Co.) Black = wax removed; gray = normal wax.

2.4.1.3 Tissue Thickness

The thickness of various plant tissues also determines the degree of resistance in some crop cultivars (Table 2.3). The foliar toughness of several cruciferous-crops adversely affects the feeding behavior of mustard beetles (Tanton 1962). Cultivars of sugarcane with thick layers of leaf epidermis and parenchyma cells deter feeding by the top shoot borer, *Scirpophaga nivella* (F.) (Chang & Shih 1959). Resistance in sorhum to the sorghum shootfly is also related to thickened cells that surround the vascular bundles of leaves (Blum 1968).

Stems thickened by increased layers of epidermal cells deter or limit entrance of stem damaging insects of some cultivars of rice, sugarcane, and wheat (Fiori & Dolan 1981, Patanakamjorn & Pathak 1967, Martin et al. 1975, Wallace et al. 1974). A thick cortex layer in the stems of the wild tomato, *Lycopersicon hirsutum*, also deters feeding by the potato aphid, *Macrosiphum euphorbiae* (Thomas) (Quiras et al. 1977). Physically based stem resistance in alfalfa to the potato leafhopper (Brewer et al. 1986) is illustrated in Figure 2.9. The fruiting structures of plants may also possess physical characteristics that impart antixenosis to insects. Thickened pod walls in some southern pea cultivars deter oviposition and larval feeding of the cowpea curculio, *Chalcodermus aeneus* Boheman (Fery & Cuthbert 1979). A similar phenomenon exists in soybean cultivars resistant to the soybean pod borer, *Grapholitha glicinvorella* (Matsumura) (Nishijima 1960).

Increased husk or hull tightness around the grain of some crop plants also imparts antixenosis. Tight-hulled rice cultivars are less susceptible to the

TABLE 2.3 Plant Tissue Thickness as an Antixenosis Factor

Plant	Insect	Tissue	Reference
Alfalfa	Alfalfa weevil	Stem	Fiori and Dolan 1981
Crucifers	Mustard beetle	Leaf	Tanton 1962
Rice	Stem borer	Stem	Patanakamjorn and Pathak 1967
Sorghum	Shoot fly	Leaf	Blum 1968
Southern pea	Cowpea curculio	Pod wall	Fery and Cuthbert 1979
Soybean	Pod borer	Pod wall	Nishijima 1960
Sugarcane	Shoot borer	Leaf	Chang and Shih 1959
	Sugarcane borer	Stem	Martin et al. 1975
Sunflower	Sunflower moth	Seed pericarp	Rogers and Kreitner 1983
Wheat	Stem sawfly	Stem	Wallace et al. 1974

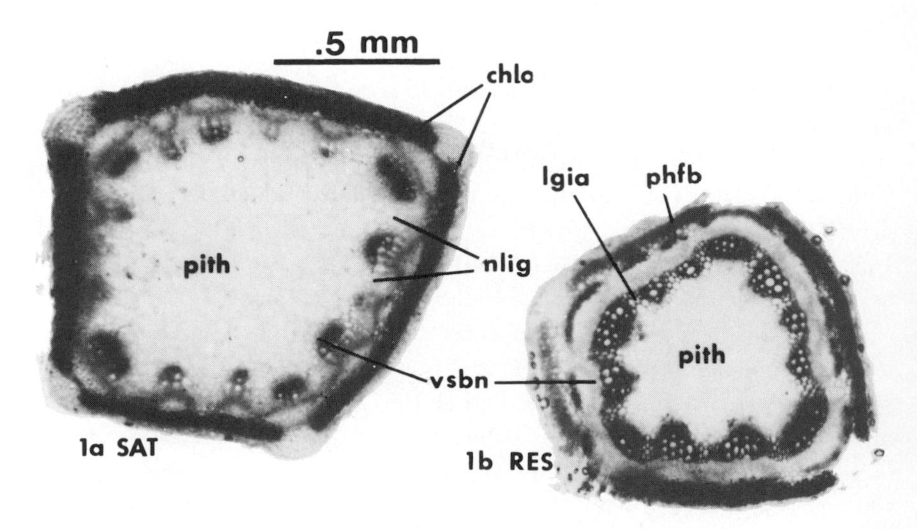

FIGURE 2.9 Physically based resistance to potato leafhopper, *Empoasca fabae* (Harris), in stems of alfalfa. (From Brewer et al. 1986. Reprinted with permission from J. Econ. Entomol., 79: 1249–1253. Copyright 1986, Entomological Society of America.) SAT = susceptible clone; RES = resistant clone; chlo = chlorophyllose tissue; nlig = nonlignified area; lgia = lignified area; phfb = phloem fibers; vsbn = vascular bundles.

Angoumois grain moth, *Sitotrogella cerealella* (Olivier), and the rice weevil, *Sitophilus oryzae* (L.), than those with loose or gaping palea and lemma (Russell & Cogburn 1977, Rosetto et al. 1973). In maize plants, tight-husked ears are resistant to penetration by the corn earworm (Wiseman et al. 1977), maize weevil, *Sitotroga zeamais* Motchulsky, and rice weevil (Singh et al. 1972, Wiseman et al. 1974). A thickened phytomelanin (armored) layer in the pericarp of sunflower seeds is responsible for the resistance of several sunflower cultivars to penetration by larvae of the sunflower moth, *Homoeosoma electellum* (Hulst) (Rogers and Kreitner 1983). Early lignification of the pericarp sclerenchymal cell wall also acts synergistically with the phytomelanin layer to increase sunflower pericarp hardness.

2.4.2 Chemical Defenses

As indicated previously, allelochemicals may act as repellents during the olfactory orientation of an insect to a resistant plant or as feeding deterrents or feeding inhibitors when an insect tastes a resistant plant. Chapman (1974)

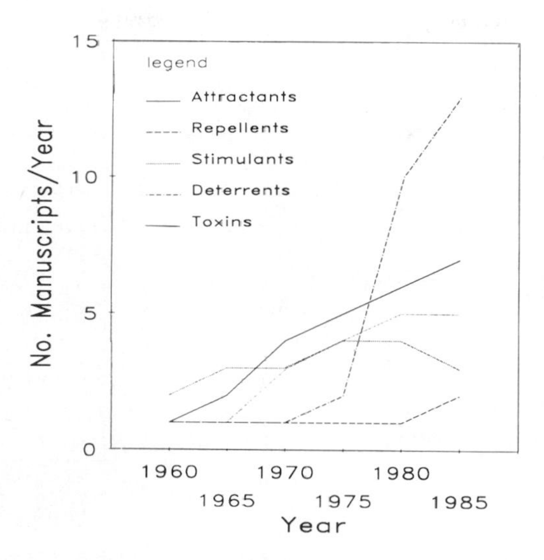

FIGURE 2.10 Types of publications dealing with the allelochemical bases of plant resistance to insects, 1955–1985.

suggested the general term *feeding inhibitor*, owing to the great diversity of the types of bioassays conducted, and the variety of terms applied to the results of the bioassay of plant allelochemicals. Schoonhoven (1982) used the term *antifeedant* to describe the function of both deterrent and inhibitory plant allelochemicals that function in antixenosis. For an in-depth discussion of antifeedants and insect sensory perception of these compounds, see Schoonhoven (1982).

Interest in the chemical bases of plant resistance to insects has increased since the mid 1960s (Fig. 2.10), with the majority of research in this area being directed at identifying insect antifeedants and feeding deterrents. Antixenosis resistance based on plant allelochemical content exists in many crops.

2.4.2.1 Repellents

Volatile hydrocarbons emitted by the foliage of resistant plants may act to repel insects (Table 2.4). Bordasch and Berryman (1977) determined that several monoterpenes from the resin vapors of the grand fir tree are repellent to the fir engraver beetle, *Scolytus ventralis* LeConte. Perttunen (1957) found a similar reaction by the bark beetles *Hylurgops palliatus* Gyll. and *Hylastes ater* Payk. Volatiles from strawberry species with high essential oil content repelfeeding by

TABLE 2.4 The Role of Repellent Allelochemicals in Antixenosis

Plant	Insect	Effect	Reference
Carrot	Carrot rust fly	Escape	Guerin and Ryan 1984
Grand fir tree	Fir engraver beetle	Repellent	Bordasch and Berryman 1977
Rice	Green rice leafhopper	Repellent	Khan and Saxena 1985
	Brown plant hopper	Repellent	Saxena and Okech 1985
	White-backed plant hopper	Repellent	Dabrowski and Rodriguez 1971
Tobacco	Green peach aphid	Escape	Johnson and Severson 1982
	Tobacco budworm	Escape	Jackson et al. 1986

the two-spotted spider mite, *Tetranychus urticae* Koch, and the strawberry spider mite, *Tetranychus turkestani* Ugarov & Nikolsik (Dabrowski & Rodriguez 1971). Similarly, volatiles from insect-resistant rice cultivars repel feeding by the green rice leafhopper, *Nephottetix virescens* Distant (Khan & Saxena 1985), the brown plant hopper, *Nilapartvata lugens* Stal (Saxena & Okech 1985), and the white-backed plant hopper, *Sogatella furcifera* (Horv.) (Khan & Saxena 1986). Resistance may also be due to the lack of perception of volatile allelochemical attractants. Guerin and Ryan (1984) determined that a decrease in the production of root volatiles by some carrot cultivars plays a role in carrot resistance to the carrot rust fly. Decreased production of diterpenes in some tobacco plant introductions also mediates resistance to the green peach aphid, *Myzus persicae* (Sulzer) (Johnson & Severson 1982), and tobacco budworm, *Heliothis virescens* (F.) (Jackson et al. 1986).

2.4.2.2 Deterrents

Deterrence of insect pests by allelochemicals exists across a broad taxonomic range of plants. The allelochemical compounds found most frequently to cause deterrence are alkaloids, flavonoids, terpene lactones, and phenols (Table 2.5). Greenbug aphids are deterred from feeding by the phenolics procyanidin, *p*-hydroxybenzalehyde, and dhurrin (Fig. 2.11) in resistant sorghum cultivars (Reese 1981, Dreyer et al. 1981). Resistance is also due to a reduction in the levels of the pectic feeding stimulant arabinogalactan, which is hydrolyzed by greenbug polysaccharases during probing and feeding (Campbell & Dreyer 1985). Sorghum cultivars resistant to greenbugs also have much greater

TABLE 2.5 Common Insect Feeding Deterrents Involved in Plant Antixenosis

Plant	Insect	Deterrent Allelochemical	Reference(s)
Alfalfa	Grass grub	Saponins	Sutherland et al. 1975
Lotus spp.	Grass grub	Isoflavan	Russell et al. 1978
Maize	European corn borer	Benzoic acids	Robinson et al. 1982
Potato	Potato leafhopper	Alkaloids	Sinden et al. 1986
Rhododendro	Root weevil	Sesquiterpene lactones	Doss et al. 1980
Rice	Stem borer	Phenolics	Das 1976
Sorghum	Greenbug	Phenolics	Reese 1981, Dreyer et al. 1981
		Pectic fructan	Campbell and Dreyer 1985
Sunflower	Sunflower moth	Sesquiterpene lactones	Gershenzon et al. 1985
Sweet clover	Blister beetles	Coumarin	Gorz et al. 1972
Wheat	Bird cherry oat aphid, English grain aphid	Phenolics	Leszcynski et al. 1985
Willow	Leaf beetle	Chlorogenic acid	Matsuda and Senbo 1986

amounts of pectic fructan, a high molecular weight polysaccharide that also inhibits greenbug feeding. Some resistant cultivars also have pectin methoxy contents that are twice as high as susceptible cultivars (Dreyer & Campbell 1983).

Phenolic acids have been implicated as a factor in the resistance of sorghum foliage to the mirgratory locust (Woodhead & Cooper-Driver 1979). *p*-Hydroxybenzaldehyde also occurs in high amounts in the surface waxes of

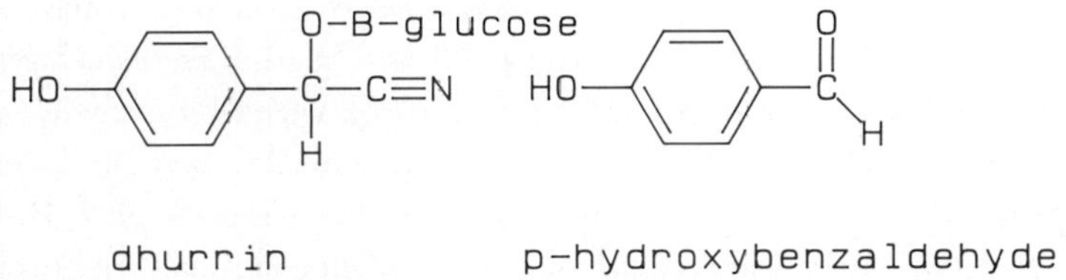

dhurrin p-hydroxybenzaldehyde

FIGURE 2.11 The phenolic insect feeding deterrents dhurrin and *p*-hydroxybenzaldehyde.

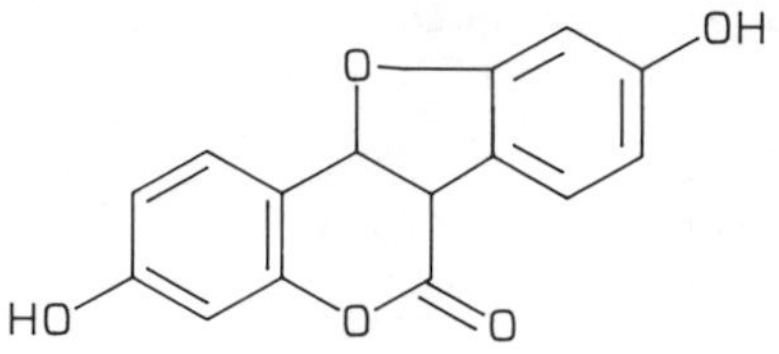

coumestrol

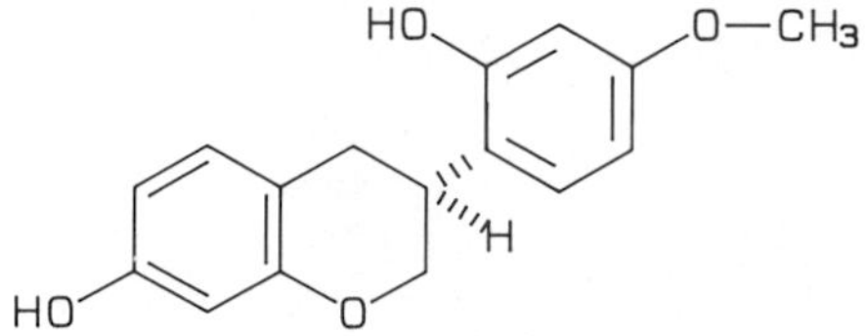

vestitol

FIGURE 2.12 Isoflavone allelochemicals from leguminous crop plants. Vestitol from alfalfa and *Lotus pedunculatus*, a feeding deterrent to the grass grub, *Costelytra zealandica* White (Russel et al. 1978, Sutherland et al. 1980); coumestrol from soybean foliage, an antibiotic compound to the soybean looper, *Pseudoplusia includens* (Walker) (Rose et al. 1988).

seedling plants (Woodhead 1982) and plays a role in sorghum resistance to locusts. In resistant wheat cultivars, dihydroxyphenols are associated with feeding deterrence to the bird cherry oat aphid and the English grain aphid, *Macrosiphum avenae* (F.) (Leszcynski et al. 1985). Phenolics also play a role in the resistance of wheat to the greenbug (Dreyer & Jones 1981) and in rice resistance to the striped stem borer, *Chilo suppressalis* (Walker) (Das 1976).

Insect feeding deterrents also occur in several forage crops. Larvae of the grass grub, *Costelytra zealandica* (White), are deterred from feeding by the isoflavone vestitol (Fig. 2.12) from *Lotus pedunculatus* (Russell et al. 1978) and by saponins from alfalfa (Sutherland 1975). Saponins, allelochemicals consisting of triterpene or a steroid linked to a sugar moiety, also act as deterrents to the Guadeloupean leaf cutter ant, *Acromyrmex octospinosis* (Reich) in some species of yam (Febvay et al. 1985). The ash gray blister beetle, *Epicauta fabricii*, the striped blister beetle, *Epicauta vittata* (F.), and the vegetable weevil, *Lisstrodes costirostris obliqus* (Klug), are deterred from feeding on the leaves of sweet clover by the phenolic compound coumarin (Fig. 2.13) (Matsumoto 1962, Gorz et al. 1972). Though coumarin is the deterring allelochemical, its occurrence depends on the hydrolysis of *trans-o*-hydroxycinnamic acid to coumarin after the enzymatic release of melilotiside in foliage tissues.

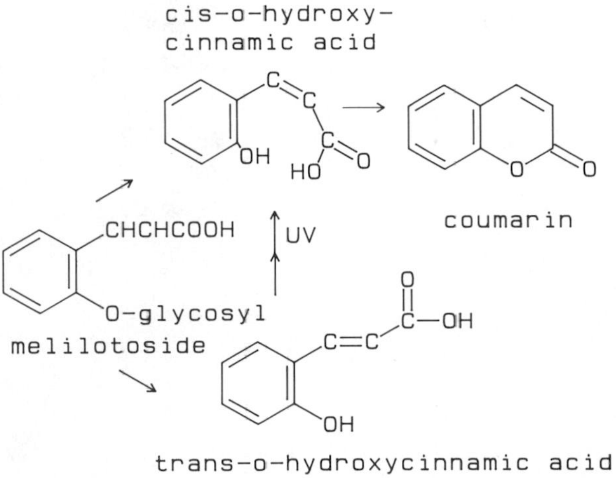

cis-o-hydroxy-
cinnamic acid

coumarin

CHCHCOOH

UV

O-glycosyl
melilotoside

trans-o-hydroxycinnamic acid

FIGURE 2.13 Coumarin, an allelochemical produced by sweet clover that deters feeding of the vegetable weevil, *Lisstroderes costirostris obliqus* Klug, the ash gray blister beetle, *Epicauta fabricii* LeConte, and the striped blister beetle *Epicauta vittata* (F.). (From Gorz et al. 1972, Matsumoto 1976.)

Foliar glycoalkaloids in wild potato species have also been noted to deter feeding of the potato leafhopper (Sinden et al. 1986). Adult pea leaf weevils, *Sitona lineatus* (L.), avoid feeding on several species of lupine with high alkaloid content (Cantot & Papineau 1983). In a similar manner, high levels of chlorogenic acid in certain species of *Salix* deter feeding of the leaf feeding beetle, *Lochmaeae capreae cribrata* (Matsuda & Senbo 1986).

An aglucone in the foliage of maize, 2,4-dihydroxy-7-methoxy-2*H*-1, 4-benzoxazin-3(4*H*)-one (DIMBOA), is one of the most widely studied plant allelochemicals affecting insect resistance. When normal, healthy maize foliage is mechanically damaged, the glucoside 2-*O*-glucosyl-4-hydroxy-1, 4-benzoxazin-3-one is enzymatically converted to DIMBOA (Fig. 2.14) (Wahlroos & Virtanen 1959). DIMBOA and its 2,-α-*O*-glucoside (DIBOA) deter feeding by the European corn borer, *Ostrinia nubilalis* (Hubner) (Robinson et al. 1982), and the greenbug (Agandona et al. 1983).

Terpene lactones deter the feeding of several different insects. The sesquiterpene lactone 8,β-sarracinoyloxycumambranolide, from the insect-resistant sunflower, *Helianthus maxmiliani*, deters feeding by the southern armyworm, *Spodoptera eridania* (Cramer); the migratory grasshopper, *Melanoplus sanguinipes* (F.); and the sunflower moth (Gershenzon et al. 1985). Insect-resistant species of rhododendron also contain high levels of the

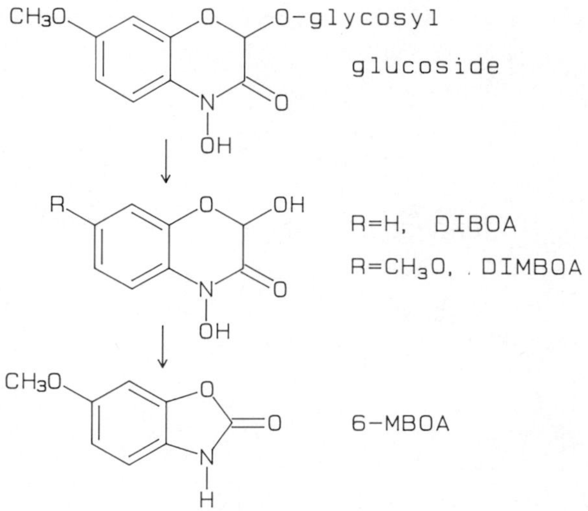

FIGURE 2.14 Production of DIMBOA [2,4-dihydroxy-8-methoxy-2*H*-1, 4-benozoxazin-3 (4*H*)-one], DIBOA (2, 4-dihydroxy-1, 4-benoxazin-3-one), and 6-MBOA (6-methoxybenzoxazolinone) by enzymatic hydroysis of mechanically damaged maize foliage.

sesquiterpene lactone germacrone, which deters feeding by the obscure root weevil, *Sciopithes obscurus* Horn (Doss et al. 1980). Norditerpene dilactones in the foliage of the insect-resistant coniferous tree, *Podocarpus gracilor*, deter feeding by the corn earworm, the fall armyworm, and the pink bollworm, *Pectinophora gossypiella* (Saunders) (Kubo et al. 1984).

Plant tannins have been considered to be insect growth inhibitors for several years, owing to their presumed action in binding with proteins to form insoluble, digestion-inhibiting complexes. Recent research (Martin et al. 1987, Martin & Martin 1982), however, indicates that there is little evidence to suggest that tannins inhibit insect digestion. The observed effects of tannins appear more likely to be due to their actions as feeding deterrents. The phenolic compounds quercitin, rutin, and procyanidin (condensed tannin) in insect-resistant cotton cultivars deter the feeding of the tobacco budworm, the corn earworm (bollworm), the pink bollworm, *Pectinophora gossypiella* (Saunders), and the two-spotted spider mite, resulting in restricted growth of these arthropods (Chan et al. 1978, Elliger et al. 1978, Lane & Schuster 1981, Lukefahr & Martin 1966). The flavonoid chrysanthemin (cyanidin-3, β-glucoside) from cotton cultivars with red floral pigmentation also deters feeding of tobacco budworm larvae (Hedin et al. 1983). Condensed tannin in the leaves of tulip and quaking aspen trees also deters feeding by the southern army worm, *Spodoptera eridania* (Cramer), and causes suppression of larval growth (Manuwoto et al. 1985).

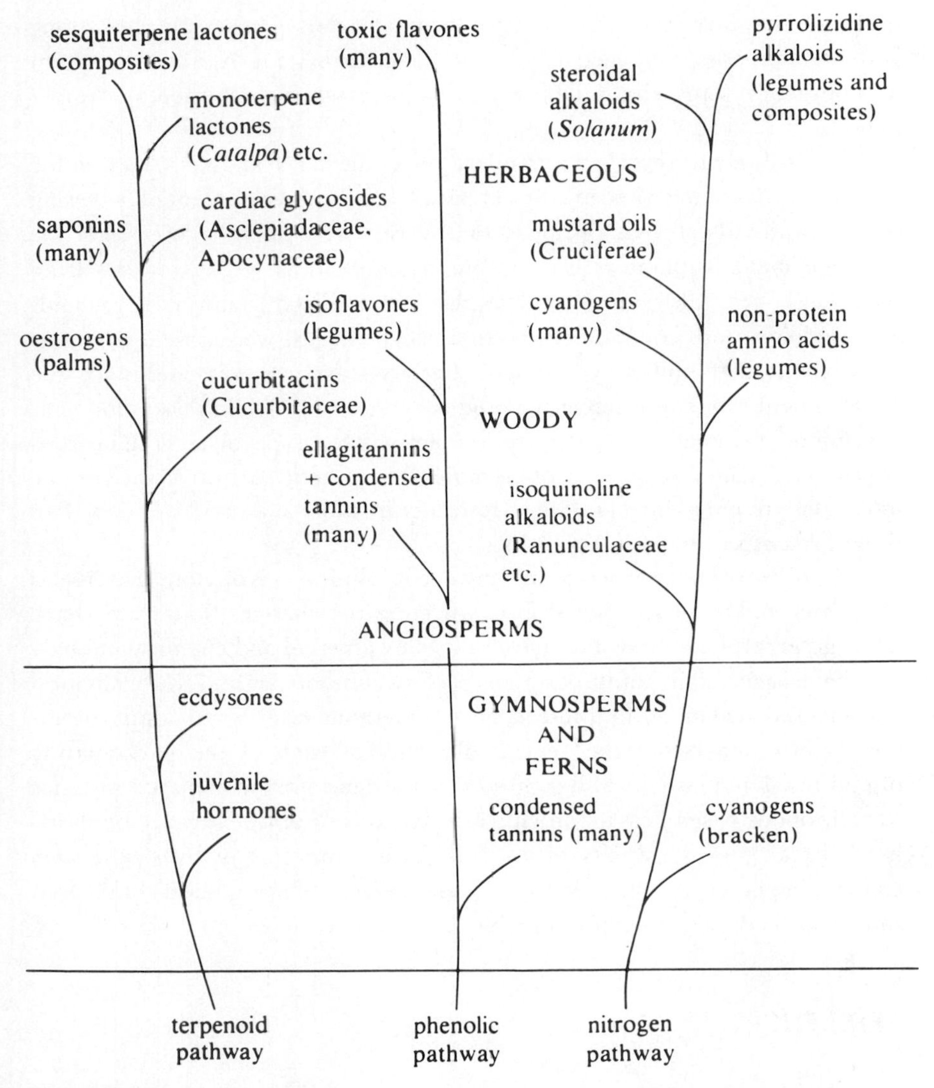

FIGURE 2.15 Evolution of insect feeding deterrents. (From Harborne 1982. Reprinted with the permission of Academic Press, Inc.)

2.5 COEVOLUTION OF PLANT DEFENSES AND INSECT BEHAVIOR

Changes in insect host plant selection depend on the dynamic equilibrium existing between insects and their potential host plants. The outcome of this equilibrium at any point in time depends on the respective genetic potentials of the plant and the insect and their rates of change relative to one another, along with the tempering influences of the external environment. At any given point in evolutionary time, either the plant or the insect will have the genetic "upper hand."

The fossil record reveals that modern insect life as we know it began in the Jurassic period of the Mesozoic Era in parallel to the development of flowering plants. Many insects living prior to the Jurassic were polyphagous. Thus the insect segment of plant–insect coevolution is often viewed as a shift from polyphagy (general feeding) to oligophagy (specialized feeding). Terpenoids and akaloids, both common classes of allelochemicals, were absent in plants until about 200 million years ago. Their occurrence coincides with the development of modern insect life, and suggests one possible reason for their development. Insect feeding deterrents themslves have coevolved with increasing chemical complexity. In all of the major biosynthetic pathways, the organic molecules present as insect feeding deterrents are more and more complex than those present in the past (Fig. 2.15).

Regardless of how we view plant–insect coevolution, it is obvious that insects have been only one selectional force affecting the changes that plants have undergone throughout evolutionary time. The physical and chemical changes that have occurred in plants in response to insect herbivory and the behavioral and metabolic changes that insects have undergone in order to adapt to new host plants underscore the genetic plasticity of each of the participating organisms. This plasticity also suggests that the development of plant resistance to arthropods based heavily on either antixenosis or antibiosis may be short-lived, because of the ability of arthropods to counteract or overcome such changes in plants (see Chapter 10). No one category of resistance is totally self-sufficient, as discussed further in Chapters 3 and 4.

REFERENCES

Ahmad, S. (ed.) 1983. *Herbivorous Insects. Host-Seeking Behavior and Mechanisms*. Academic, New York, 257 pp.

Anstey, T. H. and J. F. Moore. 1954. Inheritance of glossy foliage and cream petals in green sprouting broccoli. J. Hered. 45:39–41.

Argandona, V. H., L. J. Corcuera, H. M. Niemeyer, and B. C. Campbell. 1983. Toxicity and feeding deterrence of hydroxamic acids from gramineae in synthetic diets against the green bug, *Schizaphis graminum*. Entomol. Exp. Appl. 34:134–138.

Atkin, D. S. J. and R. J. Hamilton. 1982. The effects of plant waxes on insects. J. Nat. Prod. 45:694–696.

Baker, G. T., W. L. Parrott, and J. N. Jenkins. 1986. Sensory receptors on the larval maxillae and labia of *Heliothis zea* (Boddie) and *Heliothis virescens* (F.) (Lepidoptera:Noctuidae). Int. J. Insect Morphol. Embryol. 15:227–232.

Benedict, J. H., T. F. Leigh, and A. H. Hyer. 1983. *Lygus hesperus* (Heteroptera:Miridae) oviposition behavior, growth, and survival in relation to cotton trichome density. Environ. Entomol. 12:331–335.

Bilderback, T. E. and R. H. Mattson. 1977. Whitefly host preference associated with selected biochemical and phenotypic characteristics of poinsettias. J. Am. Soc. Hortic. Sci. 102:327–331.

Blum, A. 1968. Anatomical phenomena in seedlings of sorghum varieties resistant to the sorghum shoot fly (*Atherigona varia soccata*). Crop Sci. 8:388–390.

Bordasch, R. P. and A. A. Berryman. 1977. Host resistance of the fir engraver beetle, *Scolytus ventralis* (Coleoptera:Scolytidae). 2. Repellency of *Abies grandis* resins and some monoterpenes. Can. Entomol. 109:95–100.

Brewer, G. J., E. L. Sorensen, E. K. Horber, and G. L. Kreitner. 1986. Alfalfa stem anatomy and potato leafhopper (Homoptera:Cicadellidae) resistance. J. Econ. Entomol. 79:1249–1253.

Brues, C. F. 1920. The selection of food plants by insects with specific reference to lepidopterous larvae. Am. Nat. 54:313–22.

Burton, G. W., W. W. Hanna, J. C. Johnson, Jr., D. B. Leuck, W. G. Monson, J. B. Powell, H. D. Wells, and N. W. Widstrom. 1977. Pleiotropic effects of the tr trichomeless gene in pearl millet transpiration, forage quality and pest resistance. Crop Sci. 17:613–616.

Campbell, B. C. and D. L. Dreyer. 1985. Host-plant resistance of sorghum:Differential hydrolysis of sorghum pectic substances by polysaccharases of greenbug biotypes (*Schizaphis graminum*, Homoptera:Aphididae). Arch. Insect Biochem. Physiol. 2:203–215.

Cantot, P. and J. Papineau. 1983. Discrimination of lupine with low alkaloid content by adult *Sitona lineatus* L. Agronomie 3:937–940.

Chan, B. G., A. C. Waiss, and M. J. Lukefahr. 1978. Condensed tannin, an antibiotic chemical from *Gossypium hirsutum* L. J. Insect Physiol. 24:113–118.

Chang, H. and C. Y. Shih. 1959. A study on the leaf mid-rib structure of sugarcane as related with resistance to the top borer (*Scirpophaga nivella* F.). Taiwan Sugar Exp.

Sta. Rep. 19:53–56 (Chinese with English summary).

Chapman, R. F. 1974. The chemical inhibition of feeding by phytophagous insects: a review. Bull. Entomol. Res. 64:339–363.

Chapman, R. F. 1988. Sensory aspects of host-plant recognition by Acridoidea: Questions associated with the multiplicity of receptors and variability of response. J. Insect. Physiol. 34:167–174.

Chapman, R. F. and W. M. Blaney. 1979. In G. A. Rosenthal and D. H. Janzen (eds.). *Herbivores: Their Interaction with Secondary Plant Metabolites*. Academic, New York, pp. 161–194.

Chiang, H. S. and D. M. Norris. 1983. Morphological and physiological parameters of soybean resistance to agromyzid beanflies. Environ. Entomol. 12:260–265.

Dabrowski, Z. T. and J. G. Rodriguez. 1971. Studies on resistance of strawberries to mites. 3. Preference and nonpreference responses of *Tetranychus urticae* and *T. turkestani* to essential oils of foliage. J. Econ. Entomol. 64:387–391.

Danielson, S. D., G. R. Manglitz, and E. L. Sorensen. 1987. Resistance of perennial glandular-haired *Medicago* species to oviposition by alfalfa weevils (Coleoptera:Curculionidae). Environ. Entomol. 16:195–197.

Das, Y. T. 1976. Some factors of resistance to *Chilo suppressalis* in rice varieties. Entomol. Exp. Appl. 20:131–134.

Dethier, V. G., L. Barton-Browne, and C. N. Smith 1960. The designation of chemicals in terms of the responses they elicit from insects. J. Econ. Entomol. 53:134–136.

Doss, R. P., R. Luthi, and B. F. Hrutfiord. 1980. Germacrone, a sesquiterpene repellent to obscure root weevil from *Rhododendron edgeworthii*. Phytochemistry 19:2379–2380.

Doss, R. P., C. H. Shanks, Jr., J. D. Chamberlain, and J. K. L. Garth. 1987. Role of leaf hairs in resistance of a clone of strawberry, *Fragaria chiloensis*, to feeding by adult black vine weevil, *Otiorhynchus sulcatus* (Coleoptera:Curculionidae). Environ. Entomol. 16:764–766.

Dreyer, D. L. and B. C. Campbell. 1983. Association of the degree of methylation of intercellular pectin with plant resistance to aphids and with induction of aphid biotypes. Expertimentia 40:224–226.

Dreyer, D. L. and K. C. Jones. 1981. Feeding deterrence of flavonoids and related phenolics towards *Schizaphis graminum* and *Myzus persicae*: Aphid feeding deterrents in wheat. Phytochemistry 20:2489–2493.

Dreyer, D. L., J. C. Reese, and K. C. Jones. 1981. Aphid feeding deterrents in sorghum. Bioassay, isolation and characterization. J. Chem. Ecol. 7:273–283.

Elliger, C. A., B. G. Chan, and A. C. Waiss, Jr. 1978. Relative toxicity of minor cotton terpenoids compared to gossypol. J. Econ. Entomol. 71:161–164.

Febvay, G., P. Bourgeois, and A. Kermarrec. 1985. Antifeedants for an attine ant, *Acromyrmex octospinosus* (Reich) (Hymenoptera-Formicidae), in yam (Dioscoreaceae) leaves. Agronomie 5:439–444.

Ferguson, S., E. L. Sorensen, and E. K. Horber. 1982. Resistance to the spotted alfalfa aphid (Homoptera:Aphididae) in glandular-haired *Medicago* species. Environ. Entomol. 11:1229–1232.

Fery, R. L. and F. P. Cuthbert, Jr. 1979. Measurement of podwall resistance to the cowpea curculio in the southern pea, *Vigna unguiculata* (L.) Walp. Hortic. Sci. 14:29–30.

Fiori, B. J. and D. W. Craig. 1987. Relationship between color intensity of leaf supernatatants from resistant and susceptible birch trees and rate of oviposition by the birch leafminer (Hymenoptera:Tenthredinidae) J. Econ. Entomol. 80:1331–1333.

Fiori, B. J. and D. D. Dolan. 1981. Field tests for *Medicago* resistance against the potato leafhopper (Homoptera:Cicadellidae). Can. Entomol. 113:1049–1053.

Fraenkel, G. 1959. Evaluation of our thoughts on secondary plant substances. Science. 129:1466–1470.

George, B. W., R. S. Seay, and D. L. Coudriet. 1977. Nectariless and pubescent characters in cotton: effect on the cabbage looper. J. Econ. Entomol. 77:267–269.

Gershenzon, J., M. Rossiter, T. J. Mabry, C. E. Rogers, M. H. Blust, and T. L. Hopkins. 1985. Insect antifeedant terpenoids in wild sunflower. A possible source of resistance to the sunflower moth. In. P. A. Hedin (ed.). *Bioregulators for Pest Control.* ACS Symposium Series 276. American Chemical Society, Washington, DC, pp. 433–446.

Gorz, H. J., F. A. Haskins, and G. R. Manglitz. 1972. Effect of coumarin and related compounds on blister beetle feeding in sweetclover. J. Econ. Entomol. 65:1632–1635.

Guerin, P. M. and M. F. Ryan. 1984. Relationship between root volatiles of some carrot cultivars and their resistance to the carrot fly, *Psila rosae.* Entomol. Exp. Appl. 38:317–324.

Guerin, P. M., E. Stadler, and H. R. Buser. 1983. Identification of host plant attractants for the carrot fly, *Psila rosae.* J. Chem. Ecol. 9:843–861.

Hanson, F. E. 1983. The behavioral and neurophysiological basis of food plant selection by lepidopterous larvae. In S. Ahmad (ed.). *Herbivorous Insects Host Seeking Behavior and Mechanisms.* Academic, New York, pp. 3–23.

Harborne, J. B. 1982. *Introduction in Ecological Biochemstry.* Academic, New York, 278 pp.

Harvey, T. L. and T. J. Martin. 1980. Effects of wheat pubescence on infestations of wheat curl mite and incidence of wheat streak mosaic. J. Econ. Entomol. 73:225–227.

Hedin, P. A., J. N. Jenkins, D. H. Collum, W. H. White, and W. L. Parrott. 1983. Multiple factors in cotton contributing to resistance to the tobacco budworm, *Heliothis virescens* F. In P. A. Hedin (ed.). *Plant Resistance to Insects.* ACS Symposium Series 208. American Chemical Society, Washington, DC, pp. 347–365.

Heftmann, E. 1975. Functions of steroids in plants. Phytochemistry 14:891–901.

Hoffman, G. D. and P. B. McEvoy. 1985. The mechanism of trichome resistance in *Anaphalis margaritacea* to the meadow spittlebug *Philaenus spumarius*. Entomol. Exp. Appl. 39:123–129.

House, H. L. 1969. Effects of different proportions of nutrients on insects. Entomol. Exp. Appl. 12:651–669.

Hoxie, R. P., S. G. Wellso, and J. A. Webster. 1975. Cereal leaf beetle response to wheat trichome length and density. Environ. Entomol. 4:365–370.

Hsiao, T. H. 1969. Chemical basis of host selection and plant resistance in oligophagous insects. Entomol. Exp. Appl. 12:777–788.

Hsiao, T. H. and G. Fraenkel. 1968. The role of secondary plant substances in the food specificity of the Colorado potato beetle. Ann. Entomol. Soc. Am. 61:485–493.

Iseley, D. 1928. The relation of leaf color and leaf size to boll weevil infestation. J. Econ. Entomol. 21:553–559.

Jackson, D. M., R. F. Severson, A. W. Johnson, and G. A. Herzog. 1986. Effects of cuticular duvane diterpenes from green tobacco leaves on tobacco budworm (Lepidoptera:Noctuidae) oviposition. J. Chem. Ecol. 12:1349–1359.

Jeffree, C. E. 1986. The cuticle, epicuticular waxes and trichomes of plants, with reference to their structure, functions and evolution. In B. E. Juniper and T. R. E. Southwood (eds.). *Insects and the Plant Surface*. Edward Arnold, London, pp. 23–64.

Johnson, A. W. and R. F. Severson. 1982. Physical and chemical leaf surface characteristics of aphid resistant and susceptible tobacco. Tob. Sci. 26:98–102.

Johnson, K. J. R., E. L. Sorensen, and E. K. Horber. 1980a. Resistance in glandular haired annual *Medicago* species to feeding by adult alfalfa weevils. Environ. Entomol. 9:133–136.

Johnson, K. J. R., E. L. Sorensen, and E. K. Horber. 1980b. Resistance of glandular haired *Medicago* species to oviposition by alfalfa weevils (*Hypera postica*). Environ. Entomol. 9:241–245.

Kamel, S. A. 1965. Relationship between leaf hairness and resistance to cotton leafworm. Emp. Cotton Grow. Rev. 42:41–48.

Kennedy, J. S. 1965. Mechanisms of host plant selection. Ann. Appl. Biol. 56:317–22.

Khan, Z. R. and R. C. Saxena. 1985. Effect of steam distillate extract of a resistant rice variety on feeding behavior of *Nephotettix virescens* (Homoptera:Cicadellidae). J. Econ. Entomol. 78:562–566.

Khan, Z. R. and R. C. Saxena. 1986. Effect of steam distillate extracts of resistant and susceptible rice cultivars on behavior of *Sogatella furcifera* (Homoptera:Delphacidae). J. Econ. Entomol. 79:928–935.

Khan, Z. R., J. T. Ward, and D. M. Norris. 1986. Role of trichomes in soybean resistance to cabbage looper, *Trichoplusia ni*. Entomol. Exp. Appl. 42:109–117.

Kogan, M. and E. E. Ortman. 1978. Antixenosis-a new term proposed to replace

Painter's "non-preference" modality of resistance. Bull. Entomol. Soc. Am. 24:175–176.

Kubo, I., T. Matsumoto, and J. A. Klocke. 1984. Multichemical resistance of the conifer *Podocarpus gracilion* (Podocarpaceae) in insect attack. Chem. Ecol. 10:547–560.

Lane, H. C. and M. F. Schuster. 1981. Condensed tannins of cotton leaves. Phytochemistry 20:425–427.

Lapointe, S. L. and W. M. Tingey. 1986. Glandular trichomes of *Solanum neocardenasii* confer resistance to green peach aphid (Homoptera:Aphididae). J. Econ. Entomol. 79:1264–1268.

Leconte, C. and E. Thibout. 1981. Attraction of the leek moth, *Acrolepropsis assectella*, in an olfactometer, by volatile allelochemical compounds found in the leek, *Allium porrum*. Entomol. Exp. Appl. 30:293–300.

Lee, Y.I. 1983. The potato leafhopper, *Empoasca fabae*, soybean pubescence, and hopperburn resistance. Ph.D. thesis, University of Illinois, Urbana.

Leszczynski, B., J. Warchol, and S. Niraz. 1985. The influence of phenolic compounds on the preference of winter wheat cultivars by cereal aphids. Insect Sci. Appl. 6:157–158.

Levin, D.A. 1973. The role of trichomes in plant defense. Quart. Rev. Biol. 48:3–15.

Lukefahr, M. J. and D. F. Martin. 1966. Cotton-plant pigments as a source of resistance to the bollworm and tobacco budworm. J. Econ. Entomol. 59:176–179.

Lukefahr, M. J., C. B. Cowan, and J. E. Houghtaling. 1970. Field evaluations of improved cotton strains resistant to the cotton fleahopper. J. Econ. Entomol. 63:1101–1103.

Lukefahr, M. J., J. E. Houghtaling, and H. M. Graham. 1971. Suppression of *Heliothis* populations with glabrous cotton strains. J. Econ. Entomol. 64:486–488.

Lupton, F. G. H. 1967. The use of resistant varieties in crop protection. World Rev. Pest-Control 6:47–58.

MacLean, P. S. and R. A. Byers. 1983. Ovipositional preferences of the alfalfa blotch leafminer (Diptera:Agromyzidae) among some simple and glandular-haired *Medicago* species. Environ. Entomol. 12:1083–1086.

Maiti, R. K. and P. T. Gibson. 1983. Trichomes in segregating generations of sorghum matings. II. Association with shootfly resistance. Crop Sci. 23:76–79.

Manuwoto, S., J. M. Scriber, M. T. Hsia, and P. Sunarjo. 1985. Antibiosis/antixenosis in tulip tree and quaking aspen leaves against the polyphagous southern armyworm, *Spodoptera eridania* Oecologia 67:1–7.

Martin, J. S. and M. M. Martin. 1982. Tannin assays in ecological studies: Lack of correlation between phenolics, proanthocyanidins and protein-precipitating constituents in mature foliage of six oak species. Oecologia 54:205–211.

Martin, G. A., C. A. Richard, and S. D. Hensley. 1975. Host resistance to *Diatraea saccharalis* (F.): Relationship of sugarcane node hardness to larval damage. Environ. Entomol. 4:687–688.

Martin, J. S., M. M. Martin, and E. A. Bernays. 1987. Failure of tannic acid to inhibit digestion or reduce digestibility of plant protein in gut fluids of insect herbivores: Implications for theories of plant defense. J. Chem. Ecol. 13:605–621.

Matsuda, K. and S. Senbo. 1986. Chlorogenic acid as a feeding deterrent for the salicaceae-feeding leaf beetle, *Lochmaeae capreae cribrata*, (Coleoptera: Chrysomelidae) and other species of leaf beetles. Appl. Entomol. Zool. 21:411–416.

Matsumoto, Y. A. 1962. A dual effect of coumarin, olfactory attraction and feeding inhibition on the vegetable weevil adult, in relation to the uneatability of sweet clover leaves. Japan J. Appl. Entomol. Zool. 6:141–149.

Matsumoto, Y. A. 1970. Volatile organic sulfur compounds as insect attractants with special reference to host selection. In D. L. Wood, R. M. Silverstein, and M. Nakajima (eds.). *Control of Insect Behavior by Natural Products*. Academic, New York, pp. 133–160.

Meredith, W. R., Jr. and M. F. Schuster. 1979. Tolerance of glabrous and pubescent cottons to tarnished plant bug. Crop Sci. 19:484–488

Mustaparta, H. 1975. Behavioral responses of the pine weevil, *Hylobius abietis* L. (Col:Curculionidae) to odors activating different groups of receptor cells. J. Comp. Physiol. 102:57–63.

Nishijima, Y. 1960. Host plant preference of the soybean pod borer, *Grapholitha glicinivorella* (Matsumura) (Lep., Encosmidae). Entomol. Exp. Appl. 3:38–47.

Painter, R. H. 1951. *Insect Resistance in Crop Plants*. University of Kansas Press, Lawrence, KS, 521 pp.

Patanakamjorn, S. and M. D. Pathak. 1967. Varietal resistance of rice to Asiatic rice borer, *Chilo suppressalis* (Lepidoptera:Crambidae), and its association with various plant characters. Ann. Entomol. Soc. Am. 60:287–292.

Peiretti, R. A., I. Amini, D. W. Weibel, K. J. Starks, and R. W. McNew. 1980. Relationship of "Bloomless" (bmbm) sorghum to greenbug resistance. Crop Sci. 20:173–176.

Perttunen, V. 1957. Reactions of two bark beetle species, *Hylurgops palliatus* Gyll. and *Hyasteaster* Payk. (Col., Scoytidae) to the terpene alpha-pinene. Ann. Entomol. Fenn. 23:101–100.

Pierce, H. D., Jr. R. S. Vernon, J. H. Borden, and A. C. Oehlschlager. 1978. Host selection by *Hylemya antiqua* (Meigen): Identification of three new attractants and oviposition stimulants. J. Chem. Ecol. 4:65–72.

Prokopy, R. J. and E. D. Owens. 1983. Visual detection of plants by herbivorous insects. Ann. Rev. Entomol. 28:337–364.

Quiras, C. F., M. A. Stevens, C. M. Rick, and M. K. Kok-Yokomi. 1977. Resistance in tomato the pink form of the potato aphid, *Macrosiphum euphorbiae* (Thomas): The role of anatomy, epidermal hairs and foliage composition. J. Am. Soc. Hortic. Sci. 102:166–171.

Reed, W. 1974. Selection of cotton varieties for resistance to insect pests in Uganda. Cotton Grow. Rev. 51:106–123.

Reese, J. C. 1981. Insect dietetics: Complexities of plant-insect interactions. In G. Bhaskaran, S. Friedman and J. G. Rodriguez (eds.). *Current Topics in Insect Endocrinology and Nutrition.* Plenum, New York, pp. 317–335.

Reinhert, J. A., T. K. Broschat, and H. M. Donselman. 1983. Resistance of *Canna* sp. to the skipper butterfly, (*Calpodes ethlius*) (Lepidoptera:Hesperidae). Environ. Entomol. 12:1829–1832.

Roberts, J. J. and J. E. Foster. 1983. Effect of leaf pubescence in wheat on the bird cherry oat aphid (Homoptera:Aphidae). J. Econ. Entomol. 76:1320–1322.

Roberts, J. J., R. L. Gallun, F. L. Patterson, and J. E. Foster. 1979. Effects of wheat leaf pubescence on the Hessian fly. J. Econ. Entomol. 72:211–214.

Robinson, S. H., D. A. Wolfenbarger, and R. H. Dilday. 1980. Antixenosis of smooth leaf cotton to the ovipositional response of tobacco budworm. Crop Sci. 20:646–649.

Robinson, J. F., J. A. Klun, W. D. Guthrie, and T. A. Brindley. 1982. European corn borer (Lepidoptera: Pyralidae) leaf feeding resistance DIMBOA bioassays. J. Kans. Entomol. Soc. 55:357–364.

Rogers, C. E. and G. L. Kreitner. 1983. Phytomelanin of sunflower achemes: a mechanism for pericarp resistance to abrasion by larvae of the sunflower moth (Lepidoptera:Pyralidae). Environ. Entomol. 12:277–285.

Rogers, C. E., J. Gershenzon, N. Ohno, T. J. Mabry, R. D. Stipanovic, and G. L. Kreitner. 1987. Terpenes of wild sunflowers (*Helianthus*): An effective mechanism against seed predation by larvae of the sunflower moth, *Homoeosoma electellum* (Lepidoptera:Pyralidae). Environ. Entomol. 16:586–592.

Rose, R. L., T. C. Sparks, and C. M. Smith. 1988. Insecticide toxicity to larvae of *Pseudoplusia includens* (Walker) and *Anticarsia gemmatalis* (Hubner) (Lepidoptera) as influenced by feeding on resistant soybean (PI 227687) leaves and coumestrol. J. Econ. Entomol. 81:1288–1294.

Rosetto, C. J., R. H. Painter, and D. Wilbur. 1973. Resistancia de variodades arroz en ensa a *Sitophilus zeamais* Motchulsky. Histochem. Latineam. 9:10–18.

Russell, M. P. and R. R. Cogburn. 1977. World Collection rice varieties: resistance to seed penetration by *Sitotroga ceralella* (Olivier) (Lepidoptera: Gelchiidae). J. Stored Prod. Res. 13:103–106.

Russell. G. B., O. R. W. Sutherland, R. F. N. Hutchins, and P. E. Christmas. 1978. Vestitol: A phytoalexin with insect feeding-deterrent activity. J. Chem. Ecol. 4:571–579.

Ryan, J. D., P. Gregory, and W. M. Tingey. 1982. Phenolic oxidase activities in glandular trichomes of *Solanum berthaultii*. Phytochemistry 21:1885–1887.

Saxena, R. C. and S. H. Okech. 1985. Role of plant volatiles in resistance of selected rice varieties to brown planthopper, *Nilaparvata lugens* (Stal) (Homptera: Delphacidae). J. Chem. Ecol. 11:1601–1616.

Schoonhoven, L. M. 1982. Biological aspects of antifeedants. Entomol. Exp. Appl. 31:57–69.

Seigler, D. and P. W. Price. 1976. Secondary compounds in plants: Primary functions. Am. Natur. 110:101–105.

Shade, R. E. and L. W. Kitch. 1983. Pea aphid (Homoptera: Aphididae) biology on glandular-haired *Medicago* species. Environ. Entomol. 12:237–240.

Shade, R. E., M. J. Doskocil, and N. P. Maxon. 1979. Potato leafhopper resistance in glandular-haired alfalfa species. Crop Sci. 19:287–289.

Shifriss, O. 1981. Do *Curcurbita* plants with silvery leaves escape virus infection? Curcurbit Gen. Coop. Rep. 4:42–45.

Sinden, S. L., L. L. Sanford, W. W. Cantelo, and K. L. Deahl. 1986. Leptine glycoalkaloids and resistance to the Colorado potato beetle (Coleoptera: Chrysomelidae) in *Solanum chacoense*. Environ. Entomol. 15:1057–1062.

Singh, K., N. S. Agarwal, and G. K. Girish. 1972. The oviposition and development of *Sitophilus oryzae* (L.) (Coleoptera: Curculionidae) in different maize hybrids and corn parts. Indian J. Entomol. 34:148–154.

Smith, C. M., J. L. Frazier, L. B. Coons, and W. E. Knight. 1976. Antennal morphology of the clover head weevil, *Hypera meles* (F.). Int. J. Insect Morphol. Embryol. 5:349–355.

Stadler, E. and H. R. Buser. 1982. Oviposition stimulants for the carrot fly in the surface wax of carrot leaves. In J. H. Visser and A. K. Minks. (eds.). *Proceedings of the 5th International Symposium on Insect–Plant Relationships, Wageningen*. Pudoc, Wageningen, Netherlands, pp. 403–403.

Stork, N. E. 1980. Role of waxblooms in preventing attachment to brassicas by the mustard beetle, *Phaedon cochleariae*. Entomol. Exp. Appl. 28:99–106.

Sutherland, O. R. W., N. D. Hood, and J. R. Hiller. 1975. Lucerne root saponins: a feeding deterrent for the grass grub *Costelytra zealandica* (Coleoptera: Scarabaeidae). N. Z. J. Zool. 2:93–100.

Sutherland, O. R. W., G. B. Russell, D. R. Biggs, and G. A. Lane. 1980. Insect feeding deterrent activity of phytoalexin isoflavonoids. Biochem. Syst. Ecol. 8:73–75.

Tanton, M. T. 1962. The effect of leaf "toughness' on the feeding of larvae of the mustard beetle, *Phaedon cochleariae* Fab. Entomol. Exp. Appl. 5:74–78.

Taylor, N. L. 1956. Pubescence inheritance and leafhopper resistance relationships in alfalfa. Agron J. 48:78–81.

Thibout, E., J. Auger, and C. Lecomte. 1982. Host plant chemicals responsible for attraction and oviposition in *Acrolepiopsis assectella*. In J. H. Visser and A. K. Minks (eds.). *Proceedings of the 5th International Symposium on Insect-Plant Relationships, Wageningen*. Pudoc, Wageningen, Netherlands, pp. 107–115.

Thiery, D. and J. H. Visser. 1987. Misleading the Colorado potato beetle with an odor blend. J. Chem. Ecol. 13:1139–1146.

Tingey, W. M. and J. E. Laubengayer. 1986. Glandular trichomes of a resistant hybrid potato alter feeding behavior of the potato leafhopper (Homoptera: Cicadellidae). J. Econ. Entomol. 79:1230–1234.

Thompson, K. F. 1963. Resistance to the cabbage aphid (*Brevicoryna brassicae*) in *Brassica* plants. Nature 198:209.

Thorsteinson, A. J. 1960. Host selection in phytophagous insects. Ann. Rev. Entomol. 5:193–218.

Verma, S. C. and P. S. Mathur. 1950. The epidermal characters of sugarcane leaf in relation to insect pests, Indian J. Agric. Sci. 20:387–389.

Visser, J. H. 1983. Differential sensory preceptions of plant compounds by insects. In P. A. Hedin (ed.). *Plant Resistance to Insects*. ACS Symposium Series 208. American Chemical Society, Washington, DC, pp. 216–243.

Visser, J. H. 1986. Host odor perception in phytophagous insects. Ann. Rev. Entomol. 31:121–44.

Visser, J. H., S. Van Straten, and T. Maarse. 1979. Isolation and identification of volatiles in the foliage of potato, *Solanum tuberosum*, a host plant of the Colorado beetle, *Leptinotarsa decemlineata*. J. Chem. Ecol. 5:11–23.

Wahlroos, V. and A. I. Virtanen. 1959. The precursors of 6 MBOA in maize and wheat plants, their isolations and some of their properties. Acta Chem. Scan. 13:1906–1908.

Wallace, L. E., F. H. McNeal, and M. A. Berg. 1974. Resistance to both *Oulema melanopus* and *Cephus cinctus* in pubescent-leaved and solid stemmed wheat selections. J. Econ. Entomol. 67:105–110.

Webster, J. A. 1975. Association of plant hairs and insect resistance. An annotated bibliography. USDAARS Misc. Publ. No. 129, 18 pp.

Weibel, D. E. and K. J. Starks. 1986. Greenbug nonpreference for bloomless sorghum. Crop. Sci. 26:1151–1153.

Wessling, W. H. 1958. Genotypic reactions to boll weevil attack Upland cottons. J. Econ. Entomol. 51:508–512.

Whittaker, R. H. 1970. The biochemical ecology of higher plants. In E. Sondheimer and J. B. Simeone (eds.). *Chemical Ecology*. Academic, New York, pp. 43–70.

Widstrom, N. W., W. W. McMillian, and B. R. Wiseman. 1979. Ovipositional preferrence of the corn earworm and the development of trichomes on two exotic corn selections. Environ. Entomol. 8:833–839.

Wilson, F. D. and B. W. George. 1986. Smoothleaf and hirsute cottons: Response to insect pests and yield in Arizona. J. Econ. Entomol. 79:229–232.

Wiseman, B. R., W. W. McMillian, and N. W. Widstrom. 1974. Techniques, accomplishments, and future potential of breeding for resistance in corn to the corn earworm, fall armyworm and maize weevil, and in sorghum to the sorghum midge. In F. G. Maxwell and F. M. Harris (eds.). *Proc. Summer Inst. Biol. Control Plant Insects Dis.* Univ. Mississippi Press, Jackson, MS, pp. 381–393.

Wiseman, B. R., W. W. McMillian, and N. W. Widstrom. 1976. Feeding of corn earworm in the laboratory on excised silks of selected corn entries with notes on *Orius insidiosus*. Fla. Entomol. 59:305–308.

Wiseman, B. R., W. W. McMillian, and N. W. Widstrom. 1977. Ear characteristics and mechanisms of resistance among selected corns to corn earworm. Fla. Entomol. 60:97–103.

Woodhead, S. 1982. *p*-Hydroxybenzaldehyde in the surface wax of sorghum: its importance in seedling resistance to acridids. Entomol. Exp. Appl. 31:296–302.

Woodhead, S. and R. F. Chapman. 1986. Insect behavior and the chemistry of plant surface waxes. In B. Juniper and R. Southwood (eds.). *Insects and the Plant Surface*. Edward Arnold, London, pp. 123–135.

Woodhead, S. and G. Cooper Draver. 1979. Phenolic acids and resistance to insect attack in *Sorghum bicolor*. Biochem. Syst. Ecol. 7:309–310.

Woodhead, S. and D. E. Padgham. 1988. The effect of plant surface characteristics on resistance of rice to the brown planthopper, *Nilaparvata lugens*. Entomol. Exp. Appl. 47:15–22.

3 Antibiosis—The Effect of Plant Resistance on Insect Biology

3.1 GENERAL

Antibiosis is the category of plant resistance to insects that describes the negative effects of a resistant plant on the biology of an insect attempting to use that plant as a host. Both chemical and morphological plant defenses mediate antibiosis, and antibiotic effects of resistant plants range from mild to lethal. Lethal effects may be acute, often affecting young larvae and eggs. The chronic effects of antibiosis lead to mortality in older larvae, prepupae, pupae, and adults, when larvae and pupae fail to pupate and eclose, respectively.

Individuals surviving the direct effects of antibiosis may also suffer the debilitating effects of reduced body size and weight, prolonged periods of development in the immature stages, and reduced fecundity as surviving adults. A discussion of the types of techniques used to identify and measure antibiosis under experimental conditions is offered in Section 6.2.1.

Antibiosis occurs because of either the presence of plant allomones or the absence of plant kairomones. Antibiotic resistant cultivars may lack the proper quantities of basic insect nutrient or contain phytochemicals that are toxic to insects. Plant trichomes such as those described in Chapter 2 also have antibiotic effects on insects. Antibiosis may also occur owing to high concentrations of structural plant substances, such as lignin and silica, that reduce insect digestion.

3.2 DEFENSES OF INSECT-RESISTANT CROP PLANTS IMPARTING ANTIBIOSIS

3.2.1 Allelochemicals

3.2.1.1 Toxins

Allelochemicals such as alkaloids, ketones, and organic acids are toxic to insects (Table 3.1). The toxic nature of alkaloids produced by plants is well documented in pharmacological studies; however, in agricultural plants, there is also evidence that alkaloids mediate resistance to insects. The glycoalkaloid content of *Solanum* species resistant to the potato leafhopper, *Empoasca fabae* (Harris), is directly correlated with hopper survival (Raman et al. 1978). Leptine glycoalkaloids in *S. chalcoence* are toxic to the Colorado potato beetle, *Leptinotarsa decemlineata* (Say) (Sinden et al. 1986). The alkaloid α-tomatine also plays a role in the resistance of wild species of potato and tomato. Elliger et al. (1981) found that the α-tomatine content of the wild tomato *Lycopersicon hirsutum* f. *glabratum* is three to four times greater than that of tomato cultivars susceptible to the tomato fruitworm, *Heliothis zea* (Boddie). α-Tomatine is also partially responsible for resistance to the Colorado potato beetle in tomato (Sinden et al. 1978) and potato (Dimock et al. 1986). However, there is no strong correlation between the level of insect resistance and α-tomatine content in tomato, due presumably to the interaction of α-tomatine with free foliar sterols (Campbell & Duffey 1979). Similarly, high levels of the phenols rutin and chlorogenic acid occur in the leaves and leaf trichome tips of *L. hirsutum* f. *glabratum*, but the lack of correlation between total phenol content and tomato

TABLE 3.1 Toxic Allelochemicals Involved in Antibiosis Plant Resistance to Arthropods

Plant	Toxin	Arthropod(s) Affected	References(s)
Carrot	Chlorogenic acid	Carrot rust fly	Cole 1985
Citrus	Linalool (terpene alcohol)	Caribbean furit fly	Greany et al. 1983
Geranium	*o*-Pentadecenyl acid, *o*-heptadecenyl acid, salicyclic acid	Two-spotted spider mite	Gerhold 1984
Lettuce	Isochlorogenic acid	Lettuce root aphid	Cole 1984
Maize	DIMBOA	European corn borer	Klun et al. 1970
		Corn leaf aphid	Long et al. 1977
		Grain aphid	Argandona et al. 1980
	DIBOA	Greenbug	Argandona et al. 1983
Solanum chalcoence	Glycoalkoloids	Potato leafhopper	Sinden et al. 1986
Sorghum	Gramine (indole alkaloid)	Greenbug	Zuniga et al. 1985
		Rhopalosiphum padi	Zuniga and Corcuera 1986
Sunflower (*Helianthus* spp.)	Diterpenes, sesquiterpene lactones	Sunflower moth	Rogers et al. 1987
Tomato, potato	α-tomatine (alkaloid)	Colorado potato beetle	Sinden et al. 1978, Dimock et al. 1986
		Tomato fruitworm	Elliger et al. 1981
Wild tomato, *Lycopersicon, hirsutumn* f. *glabratum*	2-Tridecanone	Tomato fruitworm	Dimock and Kennedy 1983
		Tobbacco hornworn	Kennedy and Yamamoto 1979
		Beet armyworm	Lin et al. 1987, Williams et al. 1980
	2-Undecanone (methyl keton)	Tomato fruitworm	Farrar and Kennedy 1987,
		Beet armyworm	Lin et al. 1987

resistance to feeding by lepidopterous larvae is due to an interaction between phenols and dietary proteins (Isman & Duffey 1982a, b).

The indole alkaloid gramine (*N*, *N*-3-dimethyldimethyl indole) is a toxin responsible for the resistance of barley cultivars to the greenbug, *Schizaphis graminum* (Rondani) (Zuniga et al. 1985), and the aphid *Rhopalosiphum padi*

R = H, 2-undecanone

R = CH_2CH_3, 2-tridecanone

FIGURE 3.1 The methyl ketone toxins 2-tridecanone and 2-undecanone produced in the glandular trichomes of the wild tomato, *Solanum hirsutum* f. *glabratum*.

(Zuniga & Corcuera 1986). However, this compound also deters the feeding of both aphids (Zuniga et al. 1988), indicating that this particular resistance is due to both categories of resistance.

The methyl ketones 2-tridecanone and 2-undecanone (Fig. 3.1) play a key role in the defense of *L. hirsutum* f. *glabratum* foliage against insect defoliation. Both are produced in vacuoles on the tip of foliar glandular trichomes (Fig. 2.7b). 2-Tridecanone, produced in much higher quantities in the foliage of resistant cultivars than susceptible cultivars, is toxic to tomato fruitworm and the tobacco hornworm, *Manduca sexta* (L.) (Dimock & Kennedy 1983, Kennedy & Yamamoto 1979, Williams et al. 1980), and at least partially responsible for resistance to the Colorado potato beetle (Kennedy & Sorenson 1985). 2-Undecanone causes mortality of tomato fruitworm larvae by inhibition of pupation, but has no effect on tobacco hornworm larvae (Farrar & Kennedy 1987, 1988). Both 2-tridecanone and 2-undecanone are toxic to the tomato fruitworm, *Keiferia lycopersicalla* (Walsingham), and the beet armyworm, *Spodoptera exigua* (Hubner) (Lin et al. 1987).

Organic acids in several insect-resistant cultivars have antibiotic effects on insects. The cyclic hydroxamic acid DIMBOA and its decomposition product 6-MBOA (Fig. 2.14) in maize foliage have antibiotic effects on the European corn borer, *Ostrinia nubilalis* (Hubner) (Klun et al. 1967, 1970), and the aphid *Metopolophium dirhodum* (Walker) (Argandona et al. 1980). DIMBOA concentrations in maize, rye, and wheat cultivars are strongly correlated with insect resistance (Argandona et al. 1981, Klun & Robinson 1969). Toxic effects are evident in greenbugs feeding on diets containing concentrations of DIMBOA or DIBOA (Fig. 2.14) similar to that in greenbug-resistant Gramineae (Zuniga et al. 1983, Argandona et al. 1983). DIMBOA is also an active component in the resistance of maize to the corn leaf aphid, *Rhopalosiphum maidis* (Fitch) (Long et al. 1977), for corn leaf aphid population levels sustained on

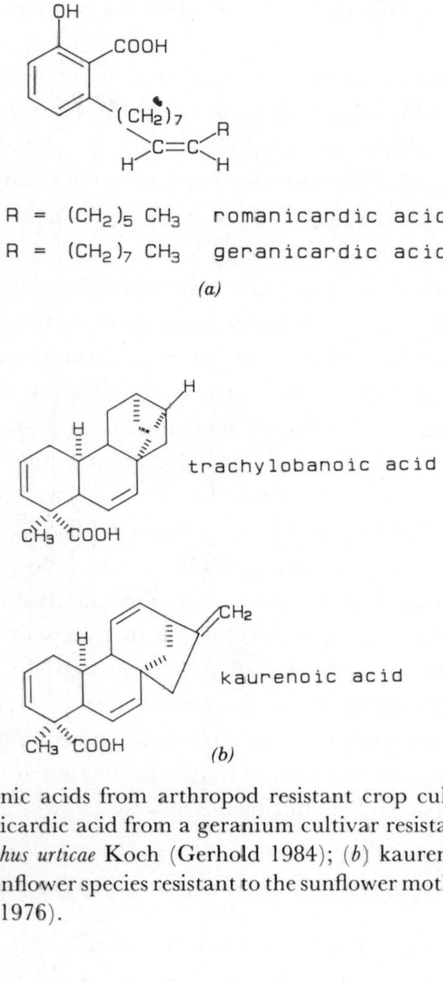

R = $(CH_2)_5\ CH_3$ romanicardic acid

R = $(CH_2)_7\ CH_3$ geranicardic acid

(a)

trachylobanoic acid

kaurenoic acid

(b)

FIGURE 3.2 Organic acids from arthropod resistant crop cultivars. (*a*) Romanicardic acid and geranicardic acid from a geranium cultivar resistant to the two spotted spider mite, *Tetranychus urticae* Koch (Gerhold 1984); (*b*) kaurenoic acid and trachylobanoic acid from sunflower species resistant to the sunflower moth, *Homeosoma electellum* Hulst (Elliger et al. 1976).

various maize cultivars are strongly correlated with the DIMBOA concentration of each cultivar (Beck et al. 1983).

Organic acids with toxic properties occur in insect-resistant cultivars of carrots and geranium. The exudate of trichomes of geranium cultivars resistant to the two-spotted spider mite, *Tetranychus urticae* Koch, consists mainly of the anicardic acid derivatives romanicardic acid and geranicardic acid (Fig. 3.2*a*). Both acids are moderately toxic to mites (Gerhold et al. 1984). The concentration of chlorogenic acid in carrot cultivars resistant to the carrot rust fly, *Psilia rosae* (F.), is closely correlated to the level of fly population development (Cole 1985). Similarly, concentrations of isochlorogenic acid in lettuce cultivars are correlated to the level of resistance to the lettuce root aphid, *Pemphigus bursarius* (L.) (Cole 1984) (see Section 6.3.1.3).

Terpene metabolites also mediate antibiosis in crop plants. Greany et al. (1983) determined that the resistance of citrus fruit to damage by the Caribbean fruit fly, *Anastrepha suspensa* (Loew), is related to the terpenoid content of the fruit rind. High mortality occurs among young larvae when they attempt to penetrate the oily layer of the citrus peel. Orange and lemon fruits, which are more resistant than grapefruit, have a much higher concentration of the terpene alcohol linalool than grapefruit, and a higher total volume of fruit peel oil glands. Several diterpenes and sesquiterpene lactones from the foliage of insect-resistant *Helianthus* species cause antibiotic symptoms in larvae of the sunflower moth, *Homoeosoma electellum* (Hulst) (Rogers et al. 1987). These symptoms include mortality, delayed development, and retarded growth. The sesquiterpene lactone 8, β-sarracinoyloxycumambranolide, produced by glandular leaf trichomes, is also toxic to larvae. Saponins, also produced in the plant terpenoid metabolic pathway (see Section 2.4.2.2.), are toxic to larvae of the grass grub, (*Costelytra zealandica* (White) feeding on resistant alfalfa and trefoil cultivars grown in the pastures of New Zealand (Sutherland et al. 1982).

A limited amount of evidence exists to indicate that plant proteins also act as insect toxins. Gatehouse et al. (1979) demonstrated that cowpea cultivars resistant to the bruchid beetle, *Callosobruchus maculatus* (F.), contain higher levels of trypsin and chymotrypsin inhibitors than susceptible cultivars. The albumin proteins of resistant cultivars that contain the trypsin inhibitor (a group of isoinhibitors) are toxic to bruchid beetle larvae.

3.2.1.2 Growth Inhibition Due to the Presence of Inhibitors

Chronic insect growth inhibition, due to either the presence of growth inhibitors or the absence of or reduction in the level of plant nutrients, is exhibited in several insect-resistant cultivars (Table 3.2). The flavone glycoside maysin (Fig. 3.3), from maize silks, is an allelochemical contained in the tassels of maize cultivars resistant to the corn earworm, *Heliothis zea* Boddie (Waiss et al. 1979). Increasing the concentration of maysin in artificial diets inhibits the growth of corn earworm larvae proportionally. However, a firm relationship between maysin concentration and penetration of maize ears by larvae in the field has not been established (Henson et al. 1984, Wiseman et al. 1985), indicating that factors other than maysin may also contribute to the resistance of maize to the corn earworm.

Coumestrol (Fig. 2.12), an isoflavone found in several legumes, displays

TABLE 3.2 Antibiotic Allelochemicals that Inhibit the Growth of Insects

Plant	Inhibitor (Allelochemical Type)	Arthropod(s) Affected	Reference
Cotton	Gossypol (sesquiterpene aldehyde)	Tobacco budworm Cotton leafworm	Lukefahr and Martin 1966 Meisner et al. 1977
	Caryophyllene oxide (terpenoid)	Tobacco budworm	Stipanovic et al. 1986
	Hemigossyppol, heliocides 1 and 2 (sesquiterpene quinones)	Tobacco budworm	Stipanovic et al. 1977
Maize	Maysin (flavone glycoside)	Corn earworm	Waiss et al. 1979
Soybean	Coumestrol (isoflavone)	Soybean looper	Rose et al. 1988, Smith 1985
Sunflower	Kaurenoic acid, trachylobanoic acid (organic acids)	Lepidoptera	Elliger et al. 1976

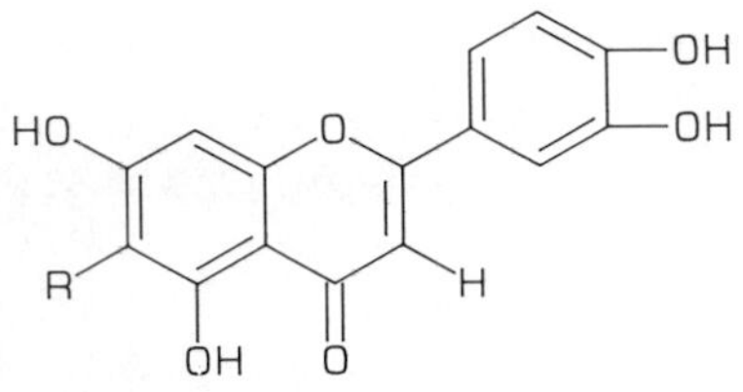

maysin

R = rhamnosyl-4-ketofucose

FIGURE 3.3 Maysin, a flavone glycoside from foliage of insect-resistant maize cultivars that inhibits growth of the corn earworm, *Heliothis zea* Boddie (Waiss et al. 1979).

pronounced estrogenic effects (Harborne 1982) in vertebrates. Coumestrol occurs in high concentration in the foliage of an insect-resistant soybean cultivar (Smith 1985). Larvae of the soybean looper, *Pseudoplusia includens* (Walker), fed diets containing coumestrol at concentrations similar to that occurring in resistant cultivars, suffer significant reductions in weight (Rose et al. 1988). The organic acids kaurenoic acid and trachylobanoic acid (Fig. 3.2*b*)

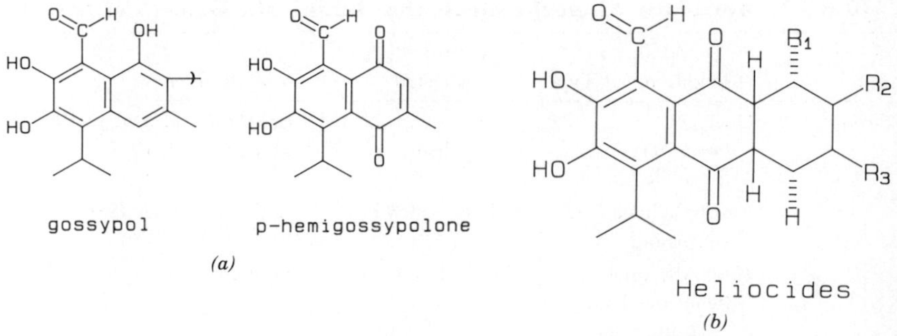

gossypol p-hemigossypolone

(a)

Heliocides

(b)

FIGURE 3.4 Terpenoid allelochemicals produced in insect-resistant cotton cultivars which inhibit the growth of foliar feeding Lepidoptera. (*a*) Gossypol and hemigossypolone; (*b*) heliocides 1 [$R_1 = CH_2CH{=}C\ (CH_3)_2$, $R_2 = CH_3$, $R_3 = H$], 2 [$R_1 = H$, $R_2 = (CH_2)_2{=}CH{=}C(CH_3)_2$, $R_3 = H$), and 3 ($R_1 = H$, $R_2 = H$, $R_3 = (CH_2)_2CH{=}C\ (CH_3)_2$] (Bell et al. 1978).

are produced in the florets of insect resistant sunflower cultivars and retard the development of larvae of several species of Lepidoptera (Elliger et al. 1976). Further evidence of their growth inhibitory properties is demonstrated by a reversal of growth inhibition when larvae are fed large quantities of cholesterol, a steroid essential to insect development.

Several terpenoids produced in the foliar pigment glands of insect-resistant cotton cultivars have antigrowth effects on several arthropod pests. The sesquiterpene aldehyde gossypol (Fig. 3.4*a*) inhibits growth in larvae of the tobacco budworm, *Heliothis virescens* (F.), (Lukefahr & Martin 1966), and the Egyptian cotton leafworm, *Spodoptera littoralis* (Boisd.) (Meisner et al. 1977). Additional terpenoid compounds, termed "x" factors for several years because of their structural complexity, are present in certain cotton cultivars in which the gossypol content does not totally explain resistance (Lukefahr et al. 1974). The "x" factors hemigossypolone (Fig. 3.4*a*), sesquiterpene quinones, and "heliocides" 1, 2, and 3 (Fig. 3.4*b*) (Bell et al. 1975, Stipanovic et al. 1977) reduce tobacco budworm growth in proportion to gossypol. A volatile monoterpene, caryophyllene oxide inhibits tobacco budworm larval growth at high concentrations and synergizes the effects of gossypol (Stipanovic et al. 1986).

3.2.1.3 Growth Inhibition Due to Reduced Levels of Nutrients

Growth inhibition in insects feeding on a crop plant may also be related to a reduction in the nutrient level of ingested food. Penny et al. (1967) determined

TABLE 3.3 Amino Acid Content of the Terminal Growth of Pea Cultivars Resistant (R) and Susceptible (S) to the Pea Aphid, *Acyrthosiphon pisum* (Harris)[a]

	Concentration (mg)[b]	
Amino Acid	'Perfection' (S)	'Laurier'(R)
L-Alanine	18.4	1.5
arginine[c]	5.4	3.1
asparagine	27.5	15.4
aspartic acid	10.3	1.9
glutamine	15.6	4.8
lysine[c]	10.8	4.8
methionine[c]	2.6	1.1
proline[c]	4.6	2.4
serine[c]	9.4	3.9

[a] From Auclair et al. (1957). Reprinted courtesy of the Entomological Society of Canada.
[b] Concentrations of all amino acids in 'Laurier' are different from those in 'Perfection' at $p < .01$
[c] Essential to pea aphid growth.

that maize plant material resistant to European corn borer larvae has an improper ascorbic acid content for adequate larval growth. Similarly, the amino acid content of the green pea cultivar 'Laurier,' which is resistant to the pea aphid, *Acyrthosiphon pisum* (Harris), is much lower than that of susceptible cultivars (Auclair et al. 1957) (Table 3.3). 'Mudgo', a rice cultivar resistant to the brown plant hopper, *Nilaparvata lugens* Stal, contains lower quantities of the amino acids asparagine and glutamic acid than susceptible rice cultivars (Sogawa & Pathak 1970) (Fig. 3.5). Research by Niraz et al. (1985) indicates that protein content and related protease activity levels of European wheat cultivars resistant to cereal aphids are substantially reduced in comparison to those of susceptible cultivars.

There is also indirect evidence that the activity of allelochemicals from resistant crop plants is related to the enzymatic activity of plant tissues. Phenylalanine ammonia-lyase, the first enzyme in the phenylpropanoid metabolism pathway, is active during herbivory at much higher levels in cultivars of several crops with insect resistance. These include lettuce resistant to the lettuce aphid (Cole 1984) (see Section 3.2.1.1), soybean cultivars resistant to the Mexican bean beetle, *Epilachna varivestis* Mulsant (Chiang et al.

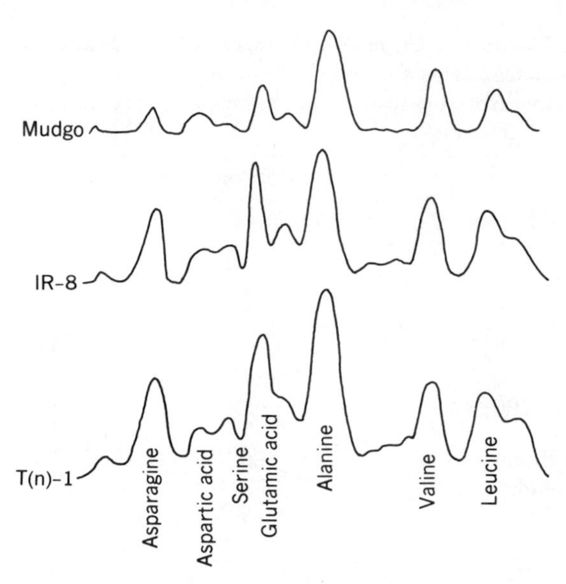

FIGURE 3.5 Asparagine and glutamic acid content of the rice cultivars 'Mudgo' (resistant), IR-8 (susceptible) and 'TN-1' (susceptible) to the brown plant hopper, *Nilaparvata lugens* Stal. (From Sogawa and Pathak 1970. Reprinted with permission of the Japan Publications Trading Co.)

1987), and cultivars of wheat resistant to a complex of aphids (Niraz et al. 1985).

3.2.2 Physical and Morphological Barriers

3.2.2.1 Hypersensitive Growth Responses of Plants

Rapidly growing plant tissues are often associated with the tolerance of crop plants to insect damage, because tissue growth may be related to the heterosis of a particular cultivar. However, rapid growth may be so dramatic that insects exhibit antibiotic effects as a result of the process of plant growth (Painter 1951). In the early 1900s, research by Hinds (1906) indicated that rapidly growing cotton boll tissues killed larvae of the boll weevil, *Anthonomus grandis grandis* Boheman (Table 3.4). More contemporary results (Adkisson et al. 1962) demonstrated similar effects on larvae of the pink bollworm. In both cases, however, the trait was linked to undesirable agronomic characters and abandoned as a breeding characteristic. Similar effects are expressed by eggplant cultivars resistant to the melon leaf miner, *Liriomyza pictella* Blanchard

TABLE 3.4 Crop Plants Whose Rapidly Growing Tissues Kill Larvae of Infesting Insects

Plant	Insect Affected	Reference
Cotton	Boll weevil	Hinds 1906
Eggplant	Melon leaf miner	Oatman 1959
Mustard	Imported cabbageworm	Shapiro and DeVay 1987
Pinus spp.	Pine shoot moth	Harris 1960
	Southern pine beetle	Hodges et al. 1979

(Oatman 1959), and in species of pine trees resistant to the pine shoot moth, *Rhyacionia buoliana* (Dennis & Schiffermuller) (Harris 1960). Hypersensitive reactions in some plants of mustard, *Brassica nigra*, to eggs of the imported cabbageworm, *Artogeia rapae* L., also occur. Shapiro and DeVay (1987) note that the leaves of some mustard plants produce a necrotic zone around the base of cabbageworm eggs, causing them to desiccate. The physical characteristics of tree oleoresin is related to insect resistance. Hodges et al. (1979) determined that longleaf pine trees and slash pine trees had higher levels of resistance than other pine species to the southern pine beetle, *Dendroctonus frontalis* Zimm., owing to higher oleoresin flow rate, viscosity, and time of crystallization. These characters all aid trees in resisting beetle invasion.

3.2.2.2 Plant Structural Factors

Insect egg, larvae, or adult mortality may also occur after contact with plant trichomes (Table 3.5). The foliage of *Solanum berthaultii*, a wild potato species, is protected by glandular trichomes from damage by the Colorado potato beetle, green peach aphid, *Myzus persicae* (Sulzer), potato leafhopper, and potato flea beetle, *Epitrix cucumeris* (Harris) (Tingey and Gibson 1978, Tingey & Sinden 1982, Gibson 1971, Casagrande 1982). The field of trichomes (Fig. 2.7*b*) is composed of tall (type B) trichomes with distal glands that exude an adhesive coating unto the tarsi of insects attempting to move about the leaf surface. As trapped insects struggle to free themselves of the type B trichome adhesive, the heads of shorter type A trichomes are ruptured (Tingey & Laubengayer 1981), releasing the two components of a natural epoxy: a resin, chlorogenic acid, and a catalyst, polyphenol oxidase (Ryan et al. 1982, 1983). Insects trapped in the hardening resin die of starvation. Foliar trichome exudates from foliage of *Solanum neocardenasii* function in a manner similar to that of *S. berthaultii* in resistance to the green peach aphid (Lapointe & Tingey 1986). Excellent

TABLE 3.5 Physical Plant Characters Mediating Antibiosis to Insects

Plant Character	Crop	Insect(s) Affected	Reference(s)
Glandular trichomes	*Solanum berthaultii*	Colorado potato beetle	Tingey and Gibson 1978
		Green peach aphid	Tingey and Sinden 1982, Gibson 1971
		Potato leafhopper, Potato flea beetle	Casagrande 1982
	Medicago spp.	Alfalfa weevil, potato leafhopper	Shade et al. 1975, 1979
	Helianthus sp.	Sunflower moth	Rogers et al. 1987
Hooked trichomes	*Phaseolus vulgaris*	Potato leafhopper	Pillemer and Tingey 1976, 1978
Simple trichomes	Wheat	Cereal leaf beetle	Wellso 1979
Silica	Maize	European corn borer	Rojanaridpiched et al. 1984
	Rice	Striped stem borer	Djamin and Pathak 1967
		African stem borer	Panda et al. 1975
		Yellow stem borer	Ukwungwu and Odebiyi 1983
	Sorghum	Sorghum shoot fly	Blum 1968
	Wheat	Hessian fly	Muller et al. 1960

reviews of the antibiotic effects of *Solanum* glandular trichomes have been prepared by Duffey (1986) and Gregory et al. (1986a, b).

Glandular trichomes contained on leaves, stems, and reproductive structures of several *Medicago* species cause antibiosis effects (Kreitner & Sorensen 1979) (Fig. 36*a*). When ruptured, these trichomes exude a sticky secretion composed of various aldehydes, alkanes, and esters (Triebe 1981). The exudate entraps and kills neonate alfalfa weevil, *Hypera postica* Gyllenhal, larvae (Shade et al. 1975) and potato leafhopper nymphs (Shade et al. 1979). Gland exudate does not entrap larger insects but does decrease adult alfalfa weevil feeding and oviposition (Johnson et al. 1980a, b) and reduce populations of the alfalfa seed chalcid, *Bruchophagus roddi* (Gussakovsky) (Brewer et al. 1983). Dense masses of simple trichomes on wild *Medicago* species also limit population development of the spotted alfalfa aphid, *Therioaphis maculata* (Buckton) (Ferguson et al. 1982).

Nonglandular plant trichomes also impart antibiotic effects. Hooked

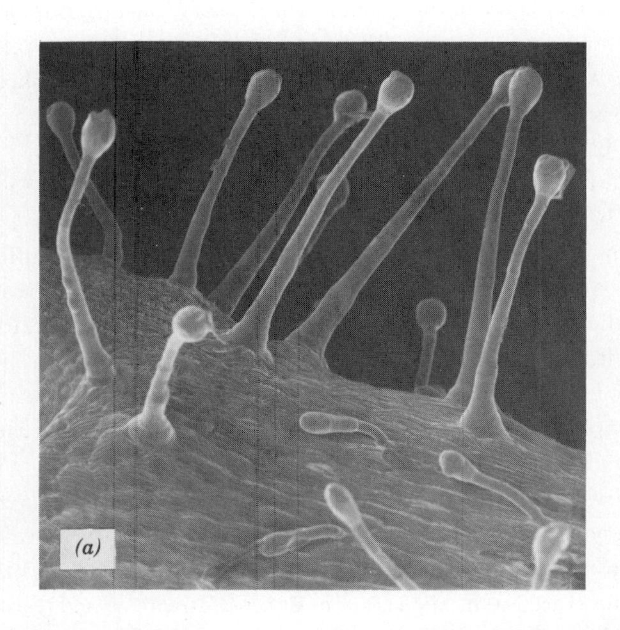

FIGURE 3.6 Plant trichomes imparting antibiosis resistance to insects. (*a*) Glandular trichomes of *Medicago disciformis* (200X). (Adapted from Kreitner & Sorensen 1979. Reprinted with the permission of the Crop Science Society of America Inc.) (*b*) Hooked trichomes of *Phaseolus vulgaris* which impale nymphs of the potato leafhopper, *Empoasca fabae* Harris (700X). (From Pillemer & Tingey 1978. Reprinted with the permission of D. W. Junk Publ. Co.).

trichomes on the foliage of the green bean, *Phaseolus vulgaris*, impale nymphs of the potato leafhopper during movement on the bean plant leaf (Fig. 3.6*b*) (Pillemer & Tingey 1976). Cultivars with a high density of hooked trichomes ($2000/cm^2$) are much more resistant than those with low trichome density ($400/cm^2$) (Pillemer & Tingey 1978).

Simple, erect trichomes on the foliage of wheat cultivars resistant to the cereal leaf beetle, *Oulema melanopus* (L.), also have antibiotic effects. Eggs deposited on the trichome "field" rising above the leaf surface suffer mortality due to desiccation and puncture by trichomes (Wellso 1979). Cereal leaf beetle larvae also die from punctures of the alimentary canal sustained after ingestion of trichome fragments (Wellso 1973). High densities of simple leaf trichomes on pubescent cotton cultivars also increase the mortality of tobacco budworm larvae by impairing their movement over the leaf surface and increasing their susceptibility to predation (Ramalho et al. 1984).

Other physical barriers in the plant epidermis itself may have antibiotic effects on insects. The presence of silica-containing cells in the rice plant imparts resistance to rice stalk boring Lepidoptera. Djamin & Pathak (1967) determined that increased stem silica content significantly increased resistance of rice to the striped stem borer, *Chilo suppressalis* (Walker) (Table 3.6). Larvae feeding on cultivars with high silica content actually lose mandibular teeth during the feeding process. Rice cultivars resistant to the African striped borer, *Chilo zacconius* Blesz., and the yellow rice borer, *Tryporyza incertulas* (Walker), also contain high levels of stem silica (Panda et al. 1975, Ukwungwu & Odebiyi 1985), which presumably affects larvae of these species in a similar manner. Increased silica content in some other gramineous plant cultivars also contributes to insect resistance in maize, sorghum, and wheat (Rojanaridpiched et al. 1984, Muller et al. 1960, Blum 1968).

TABLE 3.6 Differences Between Cultivars of Rice Resistant (R) and Susceptible (S) to the Striped Stem Borer, *Chilo suppressalis*, in Total Silica Content and Percentage of Larvae Boring into Stems

Rice Cultivar	SiO_2 Content (% dry weight)	% *C. suppressalis* Larvae Boring
Yabami Montakhab (R)	13.9	0.4
Sapan Kwai (S)	9.7	17.5

From Djamin and Pathak (1967). Adapted with permission from J. Econ. Entomol, 60:347–351. Copyright 1967, Entomological Society of America.

3.3 EFFECTS OF ALLELOCHEMICALS FROM INSECT-RESISTANT PLANTS ON INSECT METABOLISM

From the preceding discussion, it is evident that an increasing body of knowledge is accumulating concerning the chemical basis of insect resistance in crop plants. However, in only a few cases is the specific site of allelochemical activity within an insect's metabolic "machinery" actually known.

The effects of allelochemicals from resistant plants on insect metabolism are similar to those of allelochemicals that induce detoxication enzymes. Dowd et al. (1983) observed contrasting levels of hydrolytic esterase activity in larvae of the cabbage looper, *Trichoplusia ni* Hubner, and the soybean looper fed diets containing leaf extracts of resistant and susceptible soybean cultivars. Activity is reduced in midgut tissue of soybean looper larvae fed diets containing leaf extract from resistant cultivars compared to those fed diets containing extracts of leaves from a susceptible cultivar. Esterase activity is greater than normal in larvae of the cabbage looper, due presumably to this insect's more diverse host range. Soybean looper larvae fed diets containing the isoflavone coumestrol, an allelochemical active in the resistance of soybean cultivars to defoliation (Caballero et al. 1987), also undergo altered metabolic events. Such larvae have lower rates of hydrolysis of the pyrethroid insecticide fenvalerate than those fed control diet (Dowd et al 1986), and suffer enhanced fenvalerate toxicity (Rose et al. 1988), indicating that allelochemically based plant resistance synergizes insecticide toxicity.

Similar research (Kennedy 1984) with another species of pest Lepidoptera has yielded contrasting results. Ingestion of 2-tridecanone by tomato fruitworm larvae induces an enhanced level of tolerance in fruitworm larvae to the insecticide carbaryl. Presently, general predictions of the synergistic or antagonistic effects of allelochemicals from insect-resistant plants with conventional insecticides are difficult. These effects are influenced by activities of specific insect detoxification enzymes as well as the concentration at which the allelochemicals are expressed in the resistant plant. In one of the few studies of its type, Kennedy and Farrar (1987) compared the response of two populations of Colorado potato beetles (one resistant to fenvalerate, the other susceptible) to the effects of the methyl ketone, 2-tridecanone (see Section 3.2.1.1). The metabolic ability of insecticide-resistant beetles to detoxify fenvalerate has no effect on the toxic effects of 2-tridecanone. Both beetle populations are equally affected by this allelochemical from the resistant tomato plant.

REFERENCES

Adkisson, P. L. 1962. Cotton stocks screened for resistance to the pink bollworm, 1960–1961. Texas. Agric. Exp. Sta. Misc. Publ. 606.

Argandona, V. H., J. G. Luza, H. M. Niemeyer, and L. J. Corcuera. 1980. Role of hydroxamic acids in the resistance of cereals to aphids. Phytochemistry 19: 1665–1668.

Argandona, V. H., H. M. Niemeyer, and L. J. Corcuera. 1981. Effect of content and distribution of hydroxamic acids in wheat on infestation by the aphid *Schizaphis graminum*. Phytochemistry 20: 637–676.

Argandona, V. H., L. J. Corcuera, H. M. Niemeyer, and B. C. Campbell. 1983. Toxicity and feeding deterrency of hydroxamic acids from gramineae in synthetic diets against the green bug, *Schizaphis graminum*. Entomol. Exp. Appl. 34: 134–138.

Auclair, J. L., J. B. Maltais, and J. J. Cartier. 1957. Factors in resistance of peas to the pea aphid, *Acyrthosiphon pisum* (Harris) (Homoptera: Aphididae). II. Amino acids. Can. Entomol. 10: 457–464.

Beck, D. L., G. M. Dunn, D. G. Routley, and J. S. Bowman. 1983. Biochemical basis of resistance in corn to the corn lead aphid. Crop Sci. 23: 995–998.

Bell, A. A., R. D. Stipanovic, C. R. Howell, and P. R. Fryxell. 1975. Antimicrobial terpenoids of *Gossypium*: Hemigossypol, 6-methoxygossypol and 6-deoxyhemigossypol. Phytochemistry 14: 225–234.

Blum, A. 1968. Anatomical phenomena in seedlings of sorghum varieties resistant to the sorghum shoot fly (*Atherigona varia soccata*). Crop Sci. 8: 388–391.

Brewer, G. J., E. L. Sorensen, and E. K. Horber. 1983. Trichomes and field resistance of *Medicago* species to the alfalfa seed chalcid (Hymenoptera: Eurytomidae). Environ. Entomol. 12: 247–252.

Caballero, P., C. M. Smith, F. R. Fronczek, and N. H. Fischer. 1987. Isoflavonoids from soybean with potential insecticidal activity. J. Nat. Prod. 49: 1126–1129.

Campbell, B. C. and S. S. Duffey. 1979. Tomatine and parasitic wasps: Potential incompatibility of plant antibiosis with biological control. Science 205: 700–702.

Casagrande, R. A. 1982. Colorado potato beetle resistance in a wild potato, *Solanum berthaultii*. J. Econ. Entomol. 75: 368–372.

Chiang, H. S., D. M. Norris, A. Ciepiela, P. Shapiro, and A. Oosterwyk. 1987. Inducible versus constitutive PI227687 soybean resistance to Mexican bean beetle, *Epilachna* varivestis. J. Chem. Ecol. 13: 741–749.

Cole, R. A. 1984. Phenolic acids associated with the resistance of lettuce cultivars to the lettuce root aphid. Ann. Appl. Biol. 105: 129–145.

Cole, R. A. 1985. Relationship between the concentration of chlorogenic acid in carrot roots and the incidence of carrot fly larval damage. Ann. Appl. Biol. 106: 211–217.

Dimock, M. A. and G. G. Kennedy. 1983. The role of glandular trichomes in the resistance of *Lycopersicon hirsutum* f. *glabratum* to *Heliothis zea*. Entomol. Exp. Appl. 33: 263–268.

Dimock, M. B., S. L. LaPointe, and W. M. Tingey. 1986. *Solanum neocardensaii*: A new source of potato resistance to the Colorado potato beetle (Coleoptera: Chrysomelidae). J. Econ. Entomol. 79: 1269–1275.

Djamin, A. and M. D. Pathak. 1967. Role of silica in resistance to Asiatic borer (rice) *Chilo suppressalis* in rice varieties. J. Econ. Entomol. 60: 347–351.

Dowd, P. F., C. M. Smith, and T. C. Sparks. 1983. Influence of soybean leaf extracts on ester cleavage in cabbage and soybean loopers (Lepidoptera: Noctuidae). J. Econ. Entomol. 76: 700–703.

Dowd, P. F., R. R. Rose, C. M. Smith, and T. C. Sparks. 1986. Influence of extracts from soybean (*Glycine max* (L.) Merr.) leaves on hydrolytic and glutathione s-transferase activity in the soybean looper (*Pseudoplusia includens* (Walker)). J. Agric. Food Chem. 34: 444–447.

Duffey, S. S. 1986. Plant glandular trichomes: their partial role in defence against insects. In B. E. Juniper and T. R. E. Southwood (eds.) *Insects and the Plant Surface*. Edward Arnold, London, pp. 151–172.

Elliger, C. A., D. F. Zinkel, B. G. Chan, and A. C. Waiss, Jr. 1976. Diterpene acids as larval growth inhibitors. Experientia 32: 1364–1366.

Elliger, C. A., Y. Wong, B. G. Chan, and A. C. Waiss, Jr. 1981. Growth inhibitors in tomato (*Lycopersicon*) to tomato fruitworm (*Heliothis zea*). J. Chem. Ecol. 7: 753–758.

Farrar, Jr., R. R. and G. G. Kennedy. 1987. 2-Undecanone, a constituent of the glandular trichomes of *Lycopersicon hirsutum* f. *glabratum*: Effects on *Heliothis zea* and *Manduca sexta* growth and survival. Entomol. Exp. Appl. 43: 17–23.

Farrar, Jr., R. R. and G. G. Kennedy. 1988. 2-Undecanone a pupal mortality factor in *Heliothis zea*: sensitive larval stage and *in planta* activity in *Lycopersicon hirsutum* f. *glabratum*. Entomol. Exp. Appl. 47: 205–210.

Ferguson, S., E. L. Sorensen, and E. K. Horber. 1982. Resistance to the spotted alfalfa aphid (Homoptera: Aphididae) in glandular-haired *Medicago* species. Environ. Entomol. 11: 1229–1232.

Gatehouse, A. M. R., J. A. Gatehouse, P. Dobie, A. M. Kilminster, and D. Boulter. 1979. Biochemical basis of insect resistance in *Vigna unguiculata*. J. Sci. Food Agric. 30: 948–958.

Gerhold, D. L., R. Craig, and R. O. Mumma. 1984. Analysis of trichome exudate from mite-resistant geraniums. J. Chem. Ecol. 10: 713–722.

Gibson, R. W. 1971. Glandular hairs providing resistance to aphids in certain wild potato species. Ann. Appl. Biol. 68: 113–119.

Greany, P. D., S. C. Styer, P. L. Davis, P. E. Shaw, and D. L. Chambers. 1983. Biochemical resistance of citrus to fruit flies. Demonstration and elucidation of resistance to the Caribbean fruit fly, *Anastrepha suspensa*. Entomol. Expl. Appl. 34: 40–50.

Gregory, P., W. M. Tingey, D. A. Ave, and P. Y. Bouthyette. 1986a. Potato glandular trichomes: A physiochemical defense mechanism against insects. In M. B. Green and P. A. Hedin (eds.) *Natural Resistance of Plants to Pests, Role of Allelochemicals.* ACS Symposium Series 296. American Chemical Society, Washington, DC, pp. 160–167.

Gregory, P., D. A. Ave, P. Y. Bouthyette, and W. M. Tingey. 1986b. Insect-defensive chemistry of potato glandular trichomes. In B. E. Juniper and T. R. E. Southwood (eds.). *Insects and the Plant Surface.* Edward Arnold, London. pp. 181–191.

Harborne, J. B. 1982. *Introduction to Ecological Biochemistry*. Academic, New York, 278 pp.

Harris, P. 1960. Production of pine resin and its effects on survival of *Rhyacionia buoliana* (Schiff) Can. J. Zool. 38: 121–130.

Henson, A. R., M. S. Zuber, L. L. Darrah, D. Barry, L. B. Rabin, and A. C. Waiss. 1984. Evaluation of an antibiotic factor in maize silks as a means of corn earworm (Lepidoptera: Noctuidae) suppression. J. Econ. Entomol. 77:487–490.

Hinds, W. E. 1906. Proliferation as a factor in the natural control of the Mexican cotton boll weevil. U.S. Dept. Agr. Bur. Ent. Bull. 59. 45 pp.

Hodges, J. D., W. W. Elam, W. F. Watson, and T. E. Nebeker. 1979. Oleoresin characteristics and susceptibility of four southern pines to southern pine beetle (Coleoptera, Scolytidae) attacks. Can. Entomol. 111: 889–896.

Isman, M. B. and S. S. Duffey. 1982a. Toxicity of tomato phenolic compounds to the fruitworm, *Heliothis zea*. Entomol. Exp. Appl. 31: 370–376.

Isman, M. B. and S. S. Duffey. 1982b. Phenolic compounds in foliage of commercial tomato cultivars as growth inhibitors to the fruitworm, *Heliothis zea*. J. Am. Soc. Hortic. Sci. 107: 167–170.

Johnson, K. J. R., E. L. Sorensen, and E. K. Horber. 1980a. Resistance in glandular haired annual *Medicago* species to feeding by adult alfalfa weevils. Environ. Entomol. 9: 133–136.

Johnson, K. J. R., E. L. Sorensen, and E. K. Horber. 1980b. Resistance of glandular haired *Medicago* species to oviposition by alfalfa weevils (*Hypera postica*). Environ. Entomol. 9: 241–245.

Kennedy, G. G. 1984. 2-Tridecanone, tomatoes, and *Heliothis zea*: Potential incompatibility of plant antibiosis with insecticidal control. Entomol. Exp. Appl. 35: 305–311.

Kennedy, G. G. and R. R. Farrar, Jr. 1987. Response of insecticide-resistant and susceptible Colorado potato beetles, *Leptinotarsa decemlineata* to 2-tridecanone and resistant tomato foliage: the absence of cross resistance. Entomol. Exp. Appl. 45: 187–192.

Kennedy, G. G. and C. F. Sorenson. 1985. Role of glandular trichomes in the resistance of *Lycopersicon hirsutum* f. *glabratum* to Colorado potato beetle (Coleoptera: Chrysomelidae). J. Econ. Entomol. 78: 547–551.

Kennedy, G. G. and R. T. Yamamoto. 1979. A toxic factor causing resistance in a wild tomato to the tobacco hornworm and some other insects. Entomol. Exp. Appl. 26: 121–126.

Klun, J. A. and J. F. Robinson. 1969. Concentration of two 1,4-benzoxazinones in dent corn at various stages of development and its relation to resistance of the host plant to the European corn borer. J. Econ. Entomol. 62: 214–220.

Klun, J. A., C. L. Tipton and T. A. Brindley. 1967. 2,4-Dihydroxy-7-methoxy-1, 4-benozoxazin-3-one (DIMBOA), an active agent in the resistance of maize to the European corn borer. J. Econ. Entomol. 60: 1529–1533.

Klun, J. A., W. D. Guthrie, A. R. Hallauer, and W. A. Russell. 1970. Genetic nature of the concentration of 2,4-dihydroxy-7-methoxy 2H-1,4 benzoxazin-3 (4H)-one and resistance to the European corn borer in a diallell set of eleven maize inbreds. Crop Sci. 10: 87–90.

Kreitner, G. L. and E. L. Sorensen. 1979. Glandular trichomes on *Medicago* species. Crop Sci. 19: 380–384.

Lapointe, S. L. and W. M. Tingey. 1986. Glandular trichomes of *Solanum neocardenasii* confer resistance to green peach aphid (Homoptera: Aphididae). J. Econ. Entomol. 79: 1264–1268.

Lin, S. Y. H., J. T. Trumble, and J. Kumanoto. 1987. Activity of volatile compounds in glandular trichomes of *Lycopersicon* species against two insect herbivores. J. Chem. Ecol. 13: 837–850.

Long, B. J., G. M. Dunn, J. S. Bowman, and D. G. Routley. 1977. Relationship of hydroxamic acid content in corn and resistance to the corn leaf aphid. Crop Sci. 17: 55–58.

Lukefahr, M. J. and D. F. Martin. 1966. Cotton-plant pigments as a source of resistance to the bollworm and tobacco budworm. J. Econ. Entomol. 59: 176–179.

Lukefahr, M. J., T. N. Shaver, D. E. Cruhm, and J. E. Houghtaling. 1974. Location, transference, and recovery of growth inhibition factor present in three *Gossypium hirsutum* race stocks. Proc. Beltwide Cotton Prod. Res. Conf. pp. 93–95. National Cotton Council, Memphis, TN.

Meisner, J., M. Zur, E. Kabonci, and K. R. S. Ascher. 1977. Influence of gossypol content of leaves of different cotton strains on the development of *Spodoptera littoralis* larvae. J. Econ. Entomol. 70: 714–716.

Muller, B. S., R. J. Robinson, J. A. Johnson, E. T. Jones, and B. W. X. Ponnaiya. 1960. Studies on the relation between silica in wheat plants and resistance to Hessian fly attack. J. Econ. Entomol. 53: 945–949.

Niraz, S., B. Leszczynski, A. Ciepeila, A. Urbanska, and J. Warchol. 1985. Biochemical aspects of winter wheat resistance to aphids. Insect Sci. Appl. 6: 253–257.

Oatman, E. R. 1959. Host range studies of the melon leaf miner, *Liriomyza pictella* (Thompson). Ann. Entomol. Soc. Am. 52: 739–741.

Painter, R. H. 1951. *Insect Resistance in Crop Plants*. University of Kansas Press, Lawrence, KS, 521 pp.

Panda, N., B. Pradhan, A. P. Samalo, and P. S. P. Rao. 1975. Note on the relationship of some biochemical factors with the resistance in rice varieties to yellow rice borer. Indian J. Agric. Sci. 45: 499–501.

Penny, L. H., G. E. Scott, and W. D. Guthrie. 1967. Recurrent selection for European corn borer resistance in maize. Crop Sci. 7: 407–409.

Pillemer, E. A. and W. M. Tingey. 1976. Hooked trichomes: a physical barrier to a major agricultural pest. Science. 193: 482–484.

Pillemer, E. A. and W. M. Tingey. 1978. Hooked trichomes and resistance of *Phaseolus vulgaris* to *Empoasca fabae* (Harris). Entomol. Exp. Appl. 24: 83–94.

Ramalho, F. S., W. L. Parrott, J. N. Jenkins, and J. C. McCarty, Jr. 1984. Effects of cotton leaf trichomes on the mobility of newly hatched tobacco budworms (Lepidoptera: Noctuidae). J. Econ. Entomol. 77: 619–621.

Raman, K. V., W. M. Tingey, and P. Gregory. 1978. Potato glycoalkaloids: Effect on survival and feeding behaviour of the potato leafhopper. J. Econ. Entomol. 72: 337–341.

Rogers, C. E., J. Gershenzon, N. Ohno, T. J. Mabry, R. D. Stipanovic, and G. L. Kreitner. 1987. Terpenes of wild sunflowers (Heliothis): An effective mechanism against seed predation by larvae of the sunflower moth, *Homoeosoma electellum* (Lepidoptera: Pyralidae). Environ. Entomol. 16: 586–592.

Rojanaridpiched, C., V. E. Gracen, H. L. Everett, J. C. Coors, B. F. Pugh, and P. Bouthyette. 1984. Multiple factor resistance in maize to European corn borer. Maydica 29: 305–315.

Rose, R. L., T. C. Sparks, and C. M. Smith. 1988. Insecticide toxicity to larvae of the soybean looper and the velvetbean caterpillar (Lepidoptera: Noctuidae) as influenced by feeding on resistant soybean (PI227687) leaves and coumestrol. J. Econ. Entomol. 81: 1288–1294.

Ryan, J. D., P. Gregory, and W. M. Tingey. 1982. Phenolic oxidase activities in glandular trichomes of *Solanum berthaultii*. Phytochemistry 21: 1885–1887.

Ryan, J. D., P. Gregory, and W. M. Tingey. 1983. Glandular trichomes: Enzymic browning assays for improved selection of resistance to the green peach aphid. Am. Potato. J. 61: 861–868.

Shade, R. E., T. E. Thompson, and W. R. Campbell. 1975. An alfalfa weevil resistance mechanism detected in *Medicago*. J. Econ. Entomol. 68: 399–404.

Shade, R. E., M. J. Doskocil, and N. P. Maxon. 1979. Potato leafhopper resistance in glandular-haired alfalfa species. Crop Sci. 19: 287–289.

Shapiro, A. M. and J. E. DeVay. 1987. Hypersensitivity reaction of *Brassica nigra* L. (Cruciferae) kills eggs of *Pieris* butterflies (Lepidoptera: Pieridae). Oecologia 71: 631–632.

Sinden, S. L., J. M. Schalk, and A. K. Stoner, 1978. Effects of daylength and maturity of tomato plants on tomatine content and resistance to the Colorado potato beetle. J. Am. Soc. Horic. Sci 103: 596–600.

Sinden, S. L., L. L. Sanford, W. W. Cantelo, and K. L. Deahl. 1986. Leptine glycoalkaloids and resistance to the Colorado potato beetle (Coleoptera: Chrysomelidae) in *Solanum chalcoense*. Environ. Entomol. 15: 1057–1062.

Smith, C. M. 1985. Expression, mechanisms and chemistry of resistance in soybean, *Glycine max* L. (Merr.) to the soybean looper, *Pseudoplusia includens* (Walker). Insect Sci. Appl. 6: 243–248.

Sogawa, K. and M. D. Pathak. 1970. Mechanisms of brown planthopper resistance in Mudgo variety of rice (Hemiptera: Delphacidae). Appl. Entomol. Zool. 5: 145–158.

Stipanovic, R. D., A. A. Bell, D. H. O'Brien, and M. J. Lukefahr. 1977. Heliocide H2: An insecticidal sesquiterpenoid from cotton (*Gossypium*). Tetrahedron Lett. 6: 567–570.

Stipanovic, R. D., H. J. Williams, and L. A. Smith. (1986). Cotton terpenoid inhibition of *Heliothis virescens* development. In M. B. Green and P. A. Hedin (eds.). *Natural Resistance of Plants to Pests*. ACS Symposium Series 296. American Chemical Society, Washington, DC, pp. 79–94.

Sutherland, O. R. W., R. F. N. Hutchins, and W. F. Greenfield. 1982. Effects of lucerne saponins and *Lotus* condensed tannins on survival of grass grubs, *Costelytra zealandica*. N. Z. J. Zool. 9: 511–514.

Tingey, W. M. and R. M. Gibson. 1978. Feeding and mobility of the potato leafhopper impaired by glandular trichomes of *Solanum berthaultii* and *S. polyadenium*. J. Econ. Entomol. 71: 856–858.

Tingey, W. M. and J. E. Laubengayer. 1981. Defense against the green peach aphid and potato leafhopper by glandular trichomes of *Solanum berthaultii*, J. Econ. Entomol. 74: 721–725.

Tingey, W. M. and S. L. Sinden. 1982. Glandular pubescence, glycoalkaloid composition and resistance to the green peach aphid, potato leafhopper, and potato flea beetle in *Solanum berthaultii*. Am. Potato J. 59: 95–106.

Triebe, D. C., C. E. Meloan, and E. L. Sorensen. 1981. The chemical identification of the glandular hair exudate for *Medicago scutellata*. 27th Alfalfa Improvement Conference. ARM-NC-19. U.S. Department of Agriculture. p. 52.

Ukwungwu, M. N. and J. A. Odebiyi. 1985. Incidence of *Chilo zacconius* Bleszynski on some rice varieties in relation to plant characters. Insect Sci. Appl. 6: 653–656.

Waiss, A. C. Jr., B. G. Chan, C. A. Elliger, B. R. Wiseman, W. W. McMillian, N. W. Widstorm, M. S. Zuber, and A. J. Keaster. 1979. Maysin, a flavone glycoside from cornsilks with antibiotic activity toward corn earworm. J. Econ. Entomol. 72: 256–258.

Wellso, S. G. 1973. Cereal leaf beetle: Larval feeding, orientation, development, and survival on four small-grain cultivars in the laboratory. Ann. Entomol. Soc. Am. 66: 1201–1208.

Wellso, S. G. 1979. Cereal leaf beetle: Interaction with and ovipositional adaptation to a resistant wheat. Environ. Entomol. 8: 454–457.

Williams, W. G., G. G. Kennedy, R. T. Yamamoto, J. D. Thacker, and J. Bordner. 1980. 2-Tridecanone: A naturally occurring insecticide from the wild tomato *Lycopersicon hirsutum* f. *glabratum*. Science 207: 888–889.

Wiseman, B. R., N. W. Widstrom, W. W. McMillian, and A. C. Waiss, Jr. 1985. Relationship between maysin concentration in corn silk and corn earworm (Lepidoptera: Noctuidae) growth. J. Econ. Entomol. 78: 423–427.

Zuniga, G. E. and L. J. Corcuera. 1986. Effect of gramine in the resistance of barley seedligs to the aphid *Rhopalosiphum padi*. Entomol. Exp. Appl. 40: 259–262.

Zuniga, G. E., V. H. Argandona, H. M. Niemeyer, and L. J. Corcuera. 1983. Hydroxamic acid content in wild and cultivated Gramineae. Phytochemistry. 22: 2665–2668.

Zuniga, G. E., M. S. Salgado, and L. J. Corcuera. 1985. Role of an indole alkaloid in the resistance of barley seedlings to aphids. Phytochemistry. 24: 945–947.

Zuniga, G. E., E. M. Varanda, and L. J. Corcuera. 1988. Effect of gramine on the feeding behaviour of the aphids *Schizaphis graminum* and *Rhopalosiphum padi*. Entomol. Exp. Appl. 47: 161–165.

4 Tolerance—The Effect of Plant Growth Characteristics on Resistance to Insects

withstand
Recover
outgrow

4.1 DEFINITIONS

Plants may also be resistant to insects by possessing the ability to withstand or recover from damage caused by insect populations equal to those on susceptible cultivars. The expression of *tolerance* is determined by the inherent genetic ability of a plant to outgrow an insect infestation or to recover and add new growth after the destruction or removal of damaged tissues. From an agronomic perspective, the plants of a tolerant cultivar produce a greater yield than plants of nontolerant, susceptible cultivars. Unlike antixenosis and antibiosis, tolerance involves only plant characteristics and is not part of an insect–plant interaction. However, tolerance often occurs in combination with antibiosis and antixenosis. Because of its unique nature in plant resistance to insects, the quantitative assessment of tolerance is accomplished by using different experimental procedures from those used to study antixenosis or antibiosis. The differences in the types of techniques used to evaluate plant material for the three different categories of resistance are discussed in Section 6.2.1.

4.2 OCCURRENCE OF TOLERANCE IN CROP PLANTS

Tolerance exists in cultivars across a wide taxonomic range of plant families, including forage, fiber, grain, root, and vegetable crops (Table 4.1). The following discussion is not intended to be a comprehensive review of all literature pertaining to crop tolerance to insects. For additional information readers are referred to reviews of tolerance resistance to insects by Snelling (1941), Painter (1951), and Velusamy and Heinrichs (1986).

The most extensive research in the area of plant tolerance has been conducted with barley, sorghum, alfalfa, maize, and to a lesser extent rice. Tolerance in sorghum to the chinch bug, *Blissus leucopterous leucopterous* (Say), was first noted in the mid 1930s by Snelling and Dahms (1937). Dahms (1948) later evaluated the response of barley cultivars for tolerance to damage by the greenbug. A standardized mass seedling screening technique to evaluate small grains for resistance to greenbug damage was developed by Wood (1961) that identified several wheat genotypes with tolerance to greenbug damage. Tolerance was later detected in sorghum identified as resistant to the greenbug (Schuster & Starks 1973).

New sources of sorghum tolerance were detected to the chinch bug (Mize & Wilde 1986) and the spotted stalk borer, *Chilo partellus* Swinhoe (Dabrowski & Kidiavai 1983). Tolerance to the barley fly, *Delia flavibasis* Stein, was also identified in west African sorghum cultivars by Macharia and Mueke (1986). Tolerance to the brown planthopper exists in the rice cultivars 'Triveni' (Ho et al. 1982), 'Utri Rajipan' (Panda & Heinrichs 1983), progeny of an IR8 × Ptb20·cross (Nair et al. 1978), and in several wild rices (Jung-Tsung et al. 1986). Tolerance also exists in rice cultivars resistant to the rice water weevil (Oliver et al. 1972, Grigarick 1984) and the striped stem borer, *Chilo suppressalis* (Walker) (Das 1976).

Tolerance in maize cultivars resistant to the western corn rootworm results from greatly increased root volume compared to that of susceptible cultivars (Zuber et al. 1971). Correlations exist between the root volume ratios of insecticide treated and untreated plots, and the physical resistance to pulling of the root systems of plants in the two plots (Ortman et al. 1968) (Fig. 4.1). Appreciable improvements in yield have been noted in several rootworm tolerant maize hybrids (Branson et al. 1982, 1983). Tolerance in maize also exists to earworm by the corn earworm, (Wiseman et al. 1972), to seedling feeding by the chinch bug (Painter et al. 1935), and to stem boring by the spotted stalk borer (Dabrowski & Nyangiri 1983).

The resistance of several alfalfa cultivars to a complex of aphids also involves

TABLE 4.1 Incidence of Crop Plant Tolerance Resistance to Insects and Mites

Crop	Pest causing damage	Factor whose Increase Imparts Tolerance	Reference(s)
Alfalfa	Alfalfa weevil, pea aphid	Yield	Showalter et al. 1975
	Spotted alfalfa aphid	Yield	Kindler et al. 1971
	Blue alfalfa aphid	Size	Stern et al. 1980, Bishop 1982
Barley	Greenbug	Seedling survival	Dahms 1948
Cassava	*Mononychellus* sp. mite	Yield	Byrne et al. 1982
Cotton	Tarnished plant bug	Yield	Meredith and Laster 1975
	Spotted bollworm	Growth	Sharma and Agarwal 1984
Maize	Chinch bug	Survival	Painter et al. 1935
	Corn earworm	Yield	Wiseman et al. 1972
	Spotted stalk borer	Yield	Dabrowski and Nyangiri 1983
	Western corn rootworm	Yield	Zuber et al. 1971
Muskmelon	Melon aphid	Survival	Bohn et al. 1973
Okra	Jassid, *Amrasca biguttula*	Growth	Teli and Dalaya 1981
Rice	Brown plant hopper	Yield	Ho et al. 1982, Nair et al. 1978
	Fall armyworm	Yield	Lye and Smith 1988
	Rice water weevil	Yield	Oliver et al. 1972, Grigarick 1984
	Striped rice borer	Yield	Das 1976
Sorghum	Barley fly	Yield	Macharia and Mueke 1986
	Chinch bug	Survival	Maize and Wilde 1986
	Greenbug	Survival	Schuster and Starks 1973
	Spotted stalk borer	Yield	Dabrowski and Kidiavai 1983
Strawberry	Two-spotted spider mite	Yield	Schuster et al. 1980
Tomato	Two-spotted spider mite	Yield	Gilbert et al. 1966
Wheat	Greenbug	Survival	Wood 1961
Wheat grass	*Labops hesperius*	Yield	Hewitt 1980

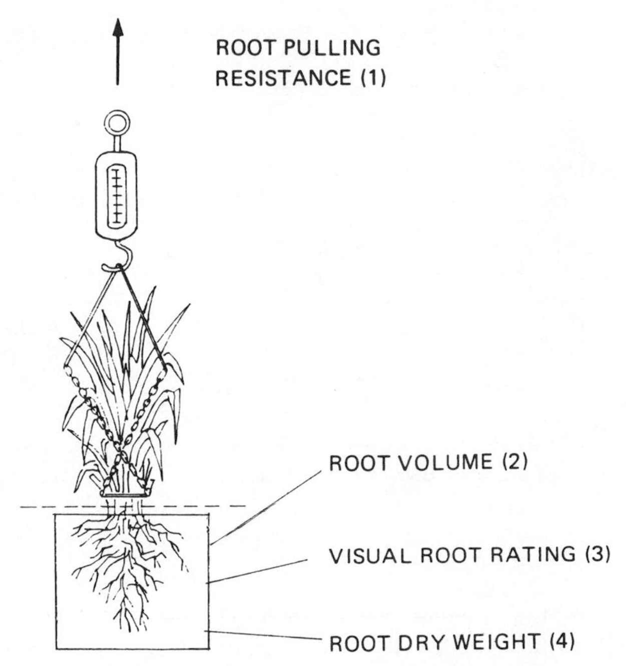

FIGURE 4.1 Measurements used in maize, rice, and sugarcane to assess tolerance to the western corn rootworm, *Diabrotica virgifera virgifera* LeConte (1, 2, 4); rice water weevil, *Lissorhoptrus oryzophilus* Kuschel (2–4); sugarcane beetle, *Euetheola humilis rugiceps* (LeConte) (1); and fall armyworm, *Spodoptera frugiperda* (J. E. Smith) (4). (From Lye and Smith 1988, Oliver et al. 1972, Ortman et al. 1968, Robinson et al. 1981, and Zuber et al. 1971.)

tolerance. The cultivars 'Dawson' and 'KS-10' tolerate damage by the pea aphid, *Acyrthrosiphon pisi* (Harris), and the spotted alfalfa aphid, *Therioaphis maculata* (Buckton), owing to increased production of dry matter, carotene, and protein (Kehr et al. 1968, Kindler et al. 1971). The newer, improved cultivars 'Lahontan' and 'Lahontan PGL' (polygenic) have tolerance to several biotypes of the spotted alfalfa aphid (Nielson & Olson 1980, Nielson & Kuehl 1982). Tolerance also exists in some high-yielding New Zealand alfalfa cultivars to an aphid complex formed by the pea aphid, spotted alfalfa aphid, and blue alfalfa aphid, *Acyrthrosiphon kondoi* Shinji (Bishop et al. 1982, Turner & Robins 1982). Plants of some alfalfa cultivars can also tolerate the effects of defoliation by larvae of the alfalfa weevil, *Hypera postica* (Gyllenhal) (Showalter et al. 1975). Tolerance to insect feeding also exists in forage grasses. Nilakhe (1987) identified tolerance to feeding by the spittlebugs *Zulia entreriana* (Berg.) and

Deois flavopicta (Stal) in the forage grasses *Andropogon gayanus* Kunth, *Bracharia brizantha* Hochst. ex. A. Rich., *Bracharia humidicola* Rendle, *Paspalum guenoarum* Archevaleta, and *Paspalum plicatulum* Michx.

Some cotton cultivars with high lint yields and the pubescent leaf character are tolerant to feeding by the tarnished plant bug (Meredith & Laster 1975, Meredith & Schuster 1979). Sharma and Agarwal (1984) determined that tolerance in cotton cultivars to stem damage from feeding by the spotted bollworm, *Earias vitella* (F.), is due to the production of greater numbers of branches in response to bollworm feeding.

Tolerance is also a component of the resistance of some cultivars of fruits and vegetables. Tolerance to the two-spotted spider mite, *Tetranychus urticae* Koch, exists in cultivars of tomatoes (Gilbert et al. 1966) and strawberries (Schuster et al. 1980). Tolerant tomato cultivars have high levels of defoliation but yields that are similar to those with little defoliation. The mite-tolerant strawberry cultivars 'Florida Belle' and 'Sequoia' have reductions in both the number and weight of fruit that are lower than those of the susceptible cultivars 'Tioga' and 'Siletz' (Schuster et al. 1980). Tolerance in some okra cultivars to feeding by the jassid *Amrasca biguttula* (Ishida) was noted by Teli and Dalaya (1981). Bohn et al. (1973) determined that tolerance in muskmelon (cantaloupe) to the melon aphid is due to the lack of leaf curling after aphid infestation. Differing levels of tolerance in plants affects the uniformity of the expression of tolerance to leaf curling.

4.3 QUANTITATIVE MEASUREMENTS OF TOLERANCE

Different techniques have been developed to evaluate the plant characteristics most commonly associated with insect tolerance. These characteristics include increases in the size and growth rate of plant leaves, stems, petioles, roots, and seed or fruit. If determinations are made in the seedling stage, plant survival is a common measurement of tolerance.

Tolerance among sorghum, wheat, barley, and rye seedlings is commonly measured by the degree of seedling survival after aphid infestation (Wood 1961). As an improved technique, Webster and Starks (1984) developed a visual rating scale to detect differences among barley seedlings for tolerance to greenbug damage. This technique involves a scale of 1 to 5, where 1 = slight seedling discoloration and 5 = severe stunting and seedling death. However, the expression of tolerance in sorghum to the greenbug may be affected by plant maturity, and Doggett et al. (1970) found that evaluations of yield differences in older, actively tillering sorghum plants are a more accurate measure of greenbug tolerance than seedling survival. Tolerance in rice to damage by the brown plant hopper, *Nilaparvata lugens* Stal, is also more accurately assessed and identified as "field resistance" in tillering vegetative plants than in seedlings (Ho et al. 1982).

Schweissing & Wilde (1979) and Panda & Heinrichs (1983) developed formulas to assess rice and sorghum tolerance to insect damage. These formulas were developed to compensate for insect weight gain during the bioassay, and are calculated as: [(mg dry weight of uninfested plants − mg dry weight of infested plants)/mg aphid dry weight]. Morgan et al. (1980) devised a further measure of sorghum tolerance in the functional plant loss index (FPLI), that combines measurements of leaf area loss and visual greenbug damage ratings. FPLI is defined as

$$1 - \left(\frac{\text{Leaf area of control} - \text{Leaf area of infested plant}}{\text{Leaf area of control}}\right) \times (1 - \text{Average visual damage rating}) \times 100$$

When insect damage is mild (in this case due to a short test duration) only leaf area is measured, using a functional plant loss (FPL) measurement where

$$\left(\frac{\text{Leaf area of uninfested control} - \text{Leaf area of infested plant}}{\text{Leaf area of uninfested control}}\right) \times 100$$

Panda & Heinrichs (1983) used a modified FPLI to determine the tolerance of rice to the brown planthopper, that was calculated as

$$1 - \left(\frac{\text{Dry wt. of infested plant}}{\text{Dry wt. of uninfested plant}}\right) \times \left(\frac{1 - \text{Damage rating}}{9}\right) \times 100$$

Measurements to identify plant material resistant to root feeding by the western corn rootworm and the rice water weevil, *Lissorhoptrus oryzophilus* Kuschel, also compensate for the differences between infested and uninfested plants. Zuber et al. (1971) compared the root volume of insecticide treated and untreated plots of maize inbreds for resistance to the western corn rootworm (Fig. 4.1). Though not defined as an FPLI, Robinson et al. (1981) evaluated resistance in rice to water weevil larval feeding using visual root ratings. Differences between the root volumes of insecticide treated and untreated rice plants were used by Oliver et al. (1972) and Cook (1988) to measure differences between cultivars in rice water weevil resistance (Fig. 4.1). Tseng et al. (1987) have developed an array of plant growth measurements to evaluate the tolerance of rice genotypes to rice water weevil larval feeding. These include seedling survival, plant height, number of leaves, number of tillers (shoots), root length, root weight, plant weight, and grain weight. Ho et al. (1982) monitored the photosynthetic rate of various rice cultivars and determined that the photosynthetic activity of the tolerant cultivar 'Triveni' was less affected after infestation by brown planthoppers than the susceptible cultivar 'Taichung Native 1.'

Regression analysis techniques have also been used to study the relationship between tolerance and antibiosis in populations of plant material evaluated for insect resistance. Lye and Smith (1988) demonstrated the partitioning of the tolerance and antibiosis resistance components to the fall armyworm, *Spodoptera frugiperda* (J. E. Smith), in rice plant introductions by plotting plant dry weight reduction against larval weight (Fig. 4.2). The intersection of a line marking the mean maximum larval weight and the regression line forms four quadrants

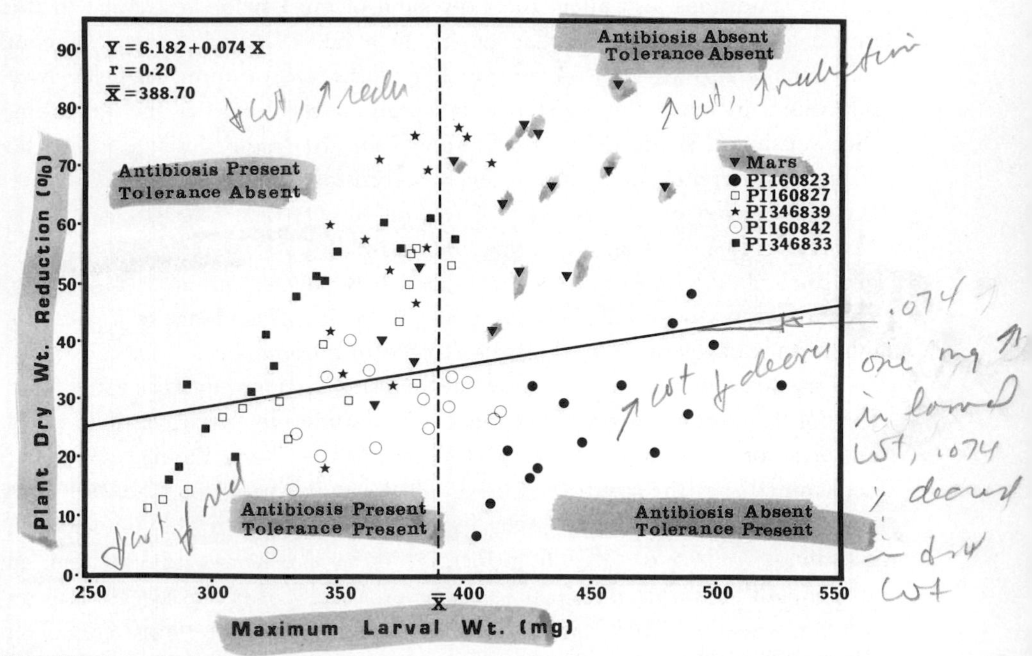

FIGURE 4.2 Tolerance to the fall armyworm, *Spodoptera frugiperda* (J. E. Smith), (plant dry weight reduction) and antibiosis (larval weight) among five rice genotypes and the susceptible cultivar 'Mars' (from Lye & Smith 1988).

that indicate different combinations of tolerance and antibiosis. The resulting scatter diagram provides an estimate of the different combinations of antibiosis and/or tolerance present in each plant introduction evaluated. A similar regression was used by Panda and Heinrichs (1983) to differentiate tolerance from antibiosis in rice cultivars resistant to the brown plant hopper and by Ortega et al. (1980) to delineate tolerance and antibiosis in maize to the fall armyworm.

4.4 FACTORS AFFECTING THE EXPRESSION OF TOLERANCE

Environmental conditions directly affect the expression of tolerance, for it is closely tied to the dynamics of plant growth. The expression of sorghum tolerance to the greenbug *Schizaphis graminum* (Rondani) is less at temperatures below ambient (15°C vs. 25°C) in the cultivar KS 30 (Schweissing & Wilde 1979). Conversely, tolerance of greenbug damage in barley, rye, and oats is greater at temperatures below ambient (Schweissing & Wilde 1978). Environmental conditions also affect the expression of muskmelon resistance to the melon aphid, *Aphis gossypii* Glover (Bohn et al. 1973). The expression of tolerance is also affected by soil nutrient levels. In sorghum, high levels of potassium increase the resistance of sorghum seedlings to the greenbug (Schweissing & Wilde 1979) (see Chapter 7 for an expanded discussion of the effects of different plant nutrient levels on resistance). Very few studies have been conducted on the genetics of tolerance. Meredith and Laster (1975) detected evidence of additive gene action in the tolerance of cotton to the tarnished plant bug, *Lygus lineolaris* (Palisot de Beauvois), but no dominance or epistatic effects were noted. Bohn et al. (1973) found evidence of a complex inheritance of tolerance to the melon aphid in muskmelon.

Very little is known about the mechanisms of plant tolerance to insect feeding. In most cases, resistance mechanisms studies involve investigations of antibiosis or antixenosis in insects. However, Maxwell and Painter (1986a, b) determined that the stunting of barley seedlings is a result of the removal of auxin during feeding by the greenbug. No auxin is noted in the honeydew of greenbugs feeding on greenbug-tolerant barley cultivars and very limited amounts of auxin are obtained from the honeydew of greenbugs feeding on tolerant wheat cultivars, compared to that obtained from greenbugs feeding on susceptible cultivars. It may be the lack of penetration of aphid stylets into stem tissues or the binding of auxins to proteins or enzyme systems in the stem tissues that prevents greenbugs from drinking auxins from tolerant cultivars.

REFERENCES

Bishop, A. L., P. J. Walters, R. H. Holtkamp, and B. C. Dominiak. 1982. Relationships between *Acyrthosiphon kondi* and damage in three varieties of alfalfa. J. Econ. Entomol. 75:118–122.

Bohn, G. W., A. N. Kishaba, J. A. Principe, and H. H. Toba. 1973. Tolerance to melon aphid in *Cucumis melo* L. J. Am. Soc. Horic. Sci. 98:37–40.

Branson, T. F., G. R. Sutter, and J. R. Fisher. 1982. Comparison of a tolerant and susceptible maize inbred under artificial infestations of *Diabrotica virgifera virgifera*: yield and adult emergence. Environ. Entomol. 11:371–372.

Branson, T. F., V. A. Welch, G. R. Sutter, and J. R. Fisher. 1983. Resistance to larvae of *Diabrotica virgifera virgifera* in three experimental maize hybrids. Environ. Entomol. 12:1509–1512.

Byrne, D. H., J. M. Guerrero, A. C. Bellotti, and V. E. Gracen. 1982. Yield and plant growth responses of *Mononychellus* mite resistant and susceptible cassava cultivars under protected vs. infested conditions. Crop Sci. 22:486–490.

Cook, C. A. 1988. Evaluation of resistance in rice, *Oryza sativa* L., to the rice water weevil, *Lissorhoptrus oryzophilus* Kuschel. M. S. Thesis. Louisiana State University. 57 pages.

Dabrowski, Z. T. and E. L. Kidiavai. 1983. Resistance of some sorghum lines to the spotted stalk-borer *Chilo partellus* under western Kenya conditions. Insect. Sci. Appl. 4:119–126.

Dabrowski, Z. T. and E. O. Nyangiri. 1983. Some field and screenhouse experiments on maize resistance to *Chilo partellus* under western Kenya conditions. Insect Sci. Appl. 4:109–118.

Dahms, R. G. 1948. Comparative tolerance of small grains to greenbugs from Oklahoma and Mississippi. J. Econ. Entomol. 41:825–826.

Das, Y. T. 1976. Cross resistance to stemborers in rice varieties. J. Econ. Entomol. 69:41–46.

Doggett, H., K. J. Starks, and S. A. Eberhart. 1970. Breeding for resistance to the sorghum shootfly. Crop Sci. 10:528–531.

Gilbert, J. C., J. T. Chinn, and J. S. Tanaka. 1966. Spider mite tolerance in multiple disease resistant tomatoes. Pro. Am. Soc. Hortic. Sci. 89:559–562.

Grigarick, A. A. 1984. General control problems with rice invertebrate pests and their control in the United States. Prot. Ecol. 7:105–114.

Hewitt, G. B. 1980. Tolerance of ten species of *Agropyron* to feeding by *Labops hesperius*. J. Econ. Entomol. 73:779–782.

Ho, D. T., E. A. Heinrichs, and F. Medrano. 1982. Tolerance of the rice variety Triveni to the brown planthopper, *Nilaparvata lugens*. Environ. Entomol. 11:598–602.

Jung-Tsung, W., E. A. Heinrichs, and F. G. Medrano. 1986. Resistance of wild rices,

Oryza spp., to the brown planthopper, *Nilaparvata lugens* (Homoptera: Delphacidae). Environ. Entomol. 15:648–653.

Kehr, W. R., G. R. Manglitz, and R. L. Ogden. 1968. Dawson alfalfa, a new variety resistant to aphids and bacterial wilt. Nebr. Agr. Exp. Sta. Bull. 497. 23 pp.

Kindler, S. D., W. R. Kehr, and R. L. Ogden. 1971. Influences of pea aphids and spotted alfalfa aphids on the stand, yield, dry matter and chemical composition of resistant and susceptible varieties of alfalfa. J. Econ. Entomol. 64:653–657.

Lye, B. H. and C. M. Smith. 1988. Evaluation of rice cultivars for antibiosis and tolerance resistance to fall armyworm (Lepidoptera: Noctuidae). Fla. Entomol. 71:254–261.

Macharia, M. and J. M. Mueke. 1986. Resistance of barley varieties to barley fly. *Delia flavibasis* Stein (Diptera: Anthomyiidae). Insect Sci. Appl. 7:75–79.

Maxwell, F. G. and R. H. Painter. 1962a. Auxin content of extracts of host plants and honeydew of different biotypes of the corn leaf aphid, *Rhopalosiphum maidis* (Fitch). J. Kans. Entomol. Soc. 35:219–233.

Maxwell, F. G. and R. H. Painter. 1962b. Auxins in honeydew of *Toxoptera graminum*, *Therioaphis maculata*, and *Macrosiphum pisi*, and their relation to degree of tolerance in host plants. Ann. Entomol. Soc. Am. 55:229–233.

Meredith, W. R., Jr., and M. L. Laster. 1975. Agronomic and genetic analysis of tarnished plant bug tolerance in cotton. Crop Sci. 15:535–538.

Meredith, W. R., Jr., and M. F. Schuster. 1979. Tolerance of glabrous and pubescent cottons to tarnished plant bug. Crop Sci. 19:484–488.

Mize, T. W. and G. Wilde. 1986. New resistant plant material to the chinch bug (Heteroptera: Lygaeidae) in grain sorghum Contribution of tolerance and antixenosis as resistance mechanisms. J. Econ. Entomol. 79:42–45.

Morgan, J., G. Wilde, and D. Johnson. 1980. Greenbug resistance in commercial sorghum hybrids in the seedling stage. J. Econ. Entomol. 73:510–514.

Nair, J. R., S. S. Nair, and N. Ramabai. 1978. A new high yielding brown planthopper tolerant variety of rice. Oryza 17:161.

Nielson, M. W. and R. O. Kuehl. 1982. Screening efficacy of spotted alfalfa aphid biotypes and genic systems for resistance in alfalfa. Environ. Entomol. 11:989–996.

Nielson, M. W. and D. L. Olson. 1982. Horizontal resistance in 'Lahontan' alfalfa to biotypes of the spotted alfalfa aphid. Environ. Entomol. 11:928–930.

Nilakhe, S. 1987. Evaluation of grasses for resistance to spittlebugs. Pesq. Agropec. Bras., Brasilia. 22:767–783.

Oliver, B. F., J. R. Gifford, and G. B. Trahan. 1972. Studies of the differences in root volume and dry root weight of rice lines where rice water weevil larvae were controlled and not controlled. Ann. Prog. Rep. Rice Res. Sta., LAES, LSU Agric. Ctr. 64:212–217.

Ortega, A., S. K. Vasal, J. Mihm, and C. Hershey. 1980. Breeding for insect resistance in maize. In F. G. Maxwell and P. R. Jennings (eds.). *Breeding Plants Resistant to Insects*. Wiley, New York, pp. 372–419.

Ortman, E. E., D. C. Peters, and P. J. Fitzgerald. 1968. Vertical-pull technique for evaluating tolerance of corn root systems to northern and western corn rootworm. J. Econ. Entomol. 61:373–375.

Painter, R. H. 1951. *Insect Resistance in Crop Plants*. University of Kansas Press, Lawerence, KS, 520 pp.

Painter, R. H., R. O. Snelling, and A. M. Brunson. 1935. Hybrid vigor and other factors in relation to chinch bug resistance in corn. J. Econ. Entomol. 28:1025–1030.

Panda, N. and E. A. Heinrichs. 1983. Levels of tolerance and antibiosis in rice varieties having moderate resistance to the brown planthopper, *Nilaparvata lugens* (Stal) (Hemiptera: Delphacidae) Environ. Entomol. 12:1204–1214.

Robinson, J. F., C. M. Smith, and G. B. Trahan. 1981. Evaluation of rice lines for rice water weevil resistance. Ann. Prog. Rep., Rice Res. Sta., LAES, LSU Agric. Ctr. 73:260–272.

Schuster, D. J. and K. J. Starks. 1973. Greenbugs: Components of host plant resistance in sorghum. J. Econ. Entomol. 66:1131–1134.

Schuster, D. J., J. F. Price, F. G. Martin, C. M. Howard, and E. E. Albregts. 1980. Tolerance of strawberry cultivars to twospotted spider mites in Florida. J. Econ. Entomol. 73:52–54.

Schweissing, F. C. and G. Wilde. 1978. Temperature influence on greenbug resistance of crops in the seedling stage. Environ. Entomol. 7:831–834.

Schweissing, F. C. and G. Wilde. 1979. Temperature and plant nutrient effects on resistance of seedling sorghum to the greenbug. J. Econ. Entomol. 72:20–23.

Sharma, H. C. and R. A. Agarwal. 1984. Factors imparting resistance to stem damage by *Earias vittela* F. (Lepidoptera: Noctuidae) in some cotton phenotypes. Prot. Ecol. 6:35–42.

Showalter, A. H., R. L. Pienkowski, and D. D. Wolf. 1975. Alfalfa weevil: host response to larval feeding. J. Econ. Entomol. 68:619–621.

Snelling, R. O. 1941. Resistance of plants to insect attack. Bot. Rev. 7:543–586.

Snelling, R. O. and R. G. Dahms. 1937. Resistant varieties of sorghum and corn in relation to chinch bug control in Oklahoma. Okla. Agric. Exp. Sta. Bull. 232.

Stern, V. M., R. Sharma, and C. Summers. 1980. Alfalfa damage from *Acrythosiphon kondoi* and economic threshold studies in southern California. J. Econ. Entomol. 73:145–148.

Teli, V. S. and V. P. Dalaya. 1981. Varietal resistance on okra to *Amrasca biguttula biguttula* (Ishida). Indian J. Agric. Sci. 51:729–731.

Tseng, S. T., C. W. Johnson, A. A. Grigarick, J. N. Rutger, and H. L. Carnahan. 1987.

Registration of short stature, early maturing, and water weevil tolerant plant material lines of rice. Crop Sci. 27:1320–1321.

Turner, J. W. and P. A. Robins. 1982. Aphid resistant lucernes. Queensland Agric. J. 108:153.

Velusamy, R. and E. A. Heinrichs. 1986. Tolerance in crop plants to insect pests. Insect Sci. Appl. 7:689–696.

Webster, J. A. and K. J. Starks. 1984. Sources of resistance in barley to two biotypes of the greenbug *Schizaphis graminum* (Rondani), Homoptera: Aphididae. Prot. Ecol. 6:51–55.

Wiseman, B. R., W. W. McMillian, and N. W. Widstrom. 1972. Tolerance as a mechanism of resistance in corn to the corn earworm. J. Econ. Entomol. 65:835–837.

Wood, E. A., Jr. 1961. Description and results of a new greenhouse technique for evaluating tolerance of small grains to the greenbug. J. Econ. Entomol. 54:303–305.

Zuber, M. S., G. J. Musick, and M. L. Fairchild. 1971. A method of evaluating corn strains for tolerance to the western corn rootworm. J. Econ. Entomol. 64:1514–1518.

SECTION II

How Is Plant Resistance to Insects Obtained?

5 Location of Sources of Plant Resistance to Insects

5.1 PROCUREMENT OF PLANT MATERIAL

Several steps are necessary to begin a program of evaluating plant material for resistance to insects. The differences between these steps are determined by the ease with which each can be accomplished. Normally, the search for resistance begins by evaluating crop cultivars grown in the geographic area where resistance is required. Sources of resistance are not randomly distributed, and this approach usually does not guarantee a high probability of identifying high levels of resistance. Frequently, a search for resistance in plant material grown outside the required location follows, requiring importation of foreign plant introductions for evaluation. Resistance may also be obtained from related species of plants, in interspecific crosses involving conventional breeding techniques. Several interspecific crosses with insect resistance have been produced. These include a *Secale* × *Triticum* cross that has yielded a *Triticale* hybrid resistant to the greenbug, and *Zea* × *Tripsacum* and *Saccharum* × *Sorghum* crosses that have resistance to diseases of monocot plants.

Normally resistance is found in a very low frequency among the plant material evaluated. Results summarized by Heinrichs (1986) from several years of evaluation for resistance to rice insect pests indicates that from 0.01 to 2.60% of the plant material evaluated was resistant. The one exception was resistance to the zigzag leafhopper; 32.9% of the 237 accessions evaluated were found to be resistant.

Searches for resistance can be directed at selection for either *allopatric* or *sympatric resistance* (plants evolving in the absence of or in the presence of insect pressure, respectively). Harris (1975) describes allopatric resistance as "the heritable plant qualities influencing the ultimate degree of damage by an insect with no previous evolutionary contact with the plant species." Conversely, sympatric resistance is that developed when insect and plant are in evolutionary contact. Leppik (1970) used disease resistance as an example and proposed that the search for insect resistance be conducted in the original home of the insect and plant. This may or may not hold true in the case of insect resistance, for in several cases, sources of insect resistance in crop plants have been obtained outside the geographic center of origin (Table 5.1). Examples include resistance in sorghum to the chinch bug, *Blissis leucopterus leucopterus* (Say); resistance in

TABLE 5.1 Examples of Host Plant Resistance Apparently Evolved in the Absence of the Insect[b]

Host	Insect(s)
Andropogon sorghum	Chinch bug
Malus sylvestris	Plum curculio, potato leafhopper
Maize	European corn borer
Pear	Pear psylla
Rubus sp.	Raspberry aphids
Rice *Oryza glaberrima*	Green rice leafhopper
Rice *O. sativa*	Rice delphacid, fall armyworm, brown plant hopper
Solanum spp.	Colorado potato beetle
Triticum spp.	Hessian fly, wheat stem sawfly
Soybean	Mexican bean beetle

[a]From Harris (1975); Jennings and Pineda (1970); Pathak (1977); and Panda and Heinrichs (1983).

wheat to the Hessian fly, *Mayetiola destructor* (Say), and wheat stem sawfly, *Cephus cinctus* Norton; resistance in soybeans to the Mexican bean beetle, *Epilachna varivestis* Mulsant; resistance in corn to the European corn borer, *Ostrinia nubilalis* (Hubner), southwestern corn borer, *Diatraea grandiosella* (Dyar), and fall armyworm *Spodoptera frugiperda* (J. E. Smith); resistance in potatoes to the Colorado potato beetle, *Leptinotarsa decemlineata* (Say); resistance in raspberries to the raspberry aphid, *Amphorophora rubi* (Kalt.); and resistance in pears to the pear psylla, *Psylla pyricola* Foerster.

One advantage to the use of allopatric resistance is that in several cases, this type of resistance is polygenic, and because it has evolved in the absence of pressure from the target insect, it may be somewhat more durable. In several instances, there has been no gene-for-gene coevolutionary progression between insect and host plant, and apparently the resistant cultivars have developed several genes that offer defense against many kinds of stresses. The real advantages of allopatric resistance over those of sympatric resistance have a largely theoretical basis, however, because both Hessian fly and raspberry aphid resistance have a monogenic basis. Additional data from the actual development of insect resistant cultivars will be required to determine more accurately the functional significance of the allopatric resistance concept.

5.2 EXISTING PLANT MATERIAL SYSTEMS

There are several sources from which to obtain crop plant material with potential resistance to insects. Agricultural scientists in many countries have developed or have access to plant material collections (Table 5.2). Efforts to obtain plant material from scientists in foreign countries should be coordinated through agencies within the supplying government. In the United States, this agency is the New Crop Introduction Branch of the United States Department of Agriculture. The Food & Agriculture Organization (FAO) of the United Nations Organization has a department concerned with obtaining seeds for use by member nations.

Many sources of plant material exist globally, including those maintained at international research centers, foreign national seed collections, and private seed companies (Table 5.3). At each location, facilities exist for receiving, processing, and storing plant material accessions. Within the United States there are several national seed collections of various crops. With the exception of the base collection at the National Seed Storage Laboratory in Fort Collins, Colorado and the Potato Collection at Sturgeon Bay, Wisconsin, most all of the

TABLE 5.2 A Directory of Agricultural Scientists Involved in Research to Develop Crop Plant Resistance to Insects

Alfalfa

J. Brewer (E), Department of Entomology, Kansas State University, Manhattan, KS 66506

E. K. Horber (E), Department of Entomology, Kansas State University, Manhattan, KS 66506

G. L. Kreitner (E), Department of Entomology, Kansas State University, Manhattan, KS 66506

G. R. Manglitz (E), Department of Entomology, University of Nebraska, Lincoln, NE 68583

R. H. Ratcliffe (E), Field Crops Laboratory, USDA-ARS, BARC-E Bldg. 467, Beltsville, MD 20705

R. E. Shade (E), Department of Entomology, Purdue University, West Lafayette, IN 47907

E. L. Sorensen (B), USDA-ARS, Department of Agronomy, Kansas State University, Manhattan, KS 66506

O. R. W. Sutherland (E), Entomology Division, Department of Scientific Research, Nelson, New Zealand

Cotton

D. W. Altman (E), USDA-ARS Cotton & Grain Crop Research, College Station, TX 77843

John Benedict (E), Texas A & M University Experiment Station, Rt 2 Box 589, Corpus Christi, TX 78410

A. J. Bockholt (E), Texas Agricultural Experiment Station, Texas A & M University, Lubbock, TX 79401

Ming-Tsang Cheo (E), Department of Plant Protection, Beijing Agricultural University People's Republic of China

H. M. Flint (B), USDA-ARS, Western Regional Cotton Research Laboratory, 4135 East Broadway Road, Phoenix, AZ 85040

J. N. Jenkins (B), USDA-ARS, Crop Science Research Laboratory, P.O. Box 5367, Mississippi State, MS 39762

H. Khalifa (E), Agricultural Research Corporation, Wad Medani, Sudan

C. Kohel (B), Southern Crops Research Laboratory, Cotton & Grain Crop Genetics, College Station, TX 77841

T. F. Leigh (E), U.S. Cotton Research Station, 17053 Shafter Ave., Shaftner, CA 93263

F. G. Maxwell (E), Department of Entomology, Texas A & M University, College Station, TX 77843

TABLE 5.2 *(Continued)*

J. C. McCarty, Jr. (B), USDA-ARS, Crop Science Research Laboratory, P.O. Box 5367, Mississippi State, MS 39762

W. L. Parrott (E), USDA-ARS, Crop Science Research Laboratory, P.O. Box 5367, Mississippi State, MS 39762

M. F. Schuster (E), Texas Agricultural Experiment Station, 17360 Cort, Dallas, TX 75252

R. D. Stipanovic (G), USDA-ARS Genetics Research, P.O. Drawer DN, College Station, TX 77841

M. F. Treacy (E), Texas Agricultural Experimental Station, Rt 2 Box 589, Corpus Christi, TX 78410

F. D. Wilson (E), USDA-ARS, Western Cotton Research Laboratory, Phoenix, AZ 85040

G. R. Zummo (B), Texas A & M Agricultural Experiment Station, Rt 2 Box 589, Corpus Christi, TX 78410

Forage Crops

L. Bush (B), Department of Agronomy, University of Kentucky, Lexington, KY 40546

R. A. Byers (E), U.S. Regional Pasture Research Laboratory, University Park, PA 16802

M. Ellsbury, USDA-ARS, Crop Science Research Laboratory, P.O. Box 5367, Mississippi State, MS 39762

K. L. Flanders (E), Department of Entomology, University of Minnesota, St. Paul, MN 55101

J. D. Hansen (range grasses) (E), USDA-ARS, Crops Research Laboratory, Logan, UT 84322

W. A. Kendall (E), USDA-ARS, Regional Pasture Laboratory, University Park, PA 16802

D. Kindler (forage grasses) (E), USDA-ARS, Plant Science Laboratory, 1301 N. Western St. Stillwater, OK 74075

J. Menn (E), USDA-ARS, BARC-W Bedg. 005, Room 232, Beltsville, MD 20705

J. J. Murray (E), Field Crops Laboratory, USDA-ARS, BARC-E Bldg. 467-C, Beltsville, MD 20705

S. S. Quisenberry (E), Department of Entomology, Louisiana State University, Baton Rouge, LA 70803

G. B. Russell (E), Ministry of Agriculture & Fisheries, Palmerston North, New Zealand

Forestry

B. J. Fiori, (E), USDA-ARS, Northeast Regional Plant Introduction Station, Geneva, NY 14456

J. W. Hanover (E), Department of Forestry, Michigan State University, East Lansing, MI 48824

Fruits

J. F. Goonewardene (apple) (E), USDA-ARS, Department of Entomology, Purdue University, West Lafayette, IN 47907

M. K. Harris (nut fruits) (E), Department of Entomology, Texas A & M University, College Station, TX 77843

C. H. Shanks (strawberries) (E), Southwest Washington Research Station, 1913 NE 78th St., Vancouver, WA 98665

Maize

M. S. Alam (E), IITA Cereal Improvement Program, PMB 5320, Ibadan, Nigeria

J. K. O. Ampofo (E), International Center for Insect Physiology and Ecology, P.O. Box 30772, Nairobi, Kenya

A. Q. Antonio (E), USDA-ARS, Department of Entomology, University of Missouri, Columbia, MO 65201

D. Barry (E), USDA-ARS, NCR, Department of Entomology, University of Missouri, Columbia, MO 65211

D. Benson (B), Department of Plant Breeding, Cornell University, Ithaca, NY 14853

N. A. Bosque-Perez (E), IITA PMB 5320, Ibadan, Nigeria

T. F. Branson (E), USDA-ARS, Northern Grain Insects Research Laboratory, RRS3, Brookings, SD 57006

J. E. Campbell (E), Pioneer International Inc., P.O. Box 85, Johnston, IA 50131

M. S. Chiang (E), Agriculture Canada Research Station, St-Jean-sur-Richelieu, Quebec J3B 628, Canada

Z. T. Dabrowski (E), International Institute of Tropical Agriculture Maize Research Program, PMB 5320, Ibadan, Nigeria

F. M. Davis (E), USDA-ARS, Crop Science Research Laboratory, P.O. Box 5367, Mississippi State, MS 39762

F. F. Dicke (E), Pioneer International Inc., P.O. Box 85, Johnston, IA 50131

V. E. Gracen (B), Seed Division, Cargill Corporation, Minneapolis, MN 55101

W. D. Guthrie (E), USDA-ARS, Corn Insects Research Unit, P.O. Box 45B Road 3, Ankeny, IA 50021

D. J. Isenhour (E), USDA-ARS, Southern Grain Insects Laboratory, Georgia Coastal Plains Experimental Station, Tifton, GA 31794

A. J. Keaster (E), 1–87 Agriculture, Department of Entomology, University of Missouri, Columbia, MO 65211

Z. B. Mayo (E), Department of Entomology, University of Nebraska, Lincoln, NE 65583

TABLE 5.2 *(Continued)*

W. W. McMillian (B), Southern Grain Insects Research Laboratory, P.O. Box 748, Tifton, GA 31793

M. E. Meehan (E), Garst Seed Company Research, P.O. Box 500, Slater, IA 50244

J. A. Mihm (E), International Maize & Wheat Improvement Center, Londres 40, APDO. Postal 6–641, Col. Juarez Deleg. Cuauhtemoc, Mexico, DF 06600 Mexico

E. E. Ortman (E), Department of Entomology, Purdue University, West Lafayette, IN 47907

J. L. Overman (E), Dekalb-Pfizer Genetics, P.O. Box 504, Union City, TN 38261

G. Rensburg (E), Grain Crops Research Unit, Private Bag X1251, Potchefstroom 2520, South Africa

J. G. Rodriguez (E), Department of Entomology, University of Kentucky, Lexington, KY 40546

J. B. Sagers (E), Northrup King Company, Hwy. 19, Stanton, MN 55081

K. N. Saxena (E), International Center for Insect Physiology and Ecology, P.O. Box 30772, Nairobi, Kenya

A. Sifuentes (E), National Institute for Agricultural Research, S. A. G. Mexico

N. W. Widstrom (B), Southern Grain Insects Research Laboratory, P.O. Box 748, Coastal Plains Experiment Station, Tifton, GA 31793

W. P. Williams (B), USDA-ARS, Crop Science Researkh Laboratory, P.O. Box 5367, Mississippi State, MS 39762

R. L. Wilson (E), USDA Plant Introduction Station, Iowa State University, Ames, IA 50011

B. R. Wiseman (E), USDA-ARS, Insect Biology and Population Management Research Laboratory, P.O. Box 748, Tifton, GA 31793

M. S. Zuber (B), USDA-ARS North Central Region, 216 Waters Hall, University of Missouri, Columbia, MO 65211

Ornamental Plants

R. P. Doss (E), Ornamental Plant Research, Laboratory, Western Washington Research & Extension Center, Puyallup, WA 98371

C. S. Koehler (E), Department of Entomology, 137 Giannini Hall, University of California, Berkeley, CA 94720

P. B. Schultz (E), Virginia Truck & Ornamental Research Station, 1444 Diamond Springs Rd., Virginia Beach, VA 23455

Physiology and Behavior

J. Auclair (E), Department of Biological Sciences CP 6128, Université de Montréal SUCCA, Montréal PQ, Canada H3C 3J7

D. Ave (E), Ecogen Inc., 2005 Cabot Blvd. West, Langhorne, PA 19047–1810

E. A. Bernays (E), Department of Entomology, University of California, Berkeley, CA 94720

T. Jermy (E), Research Institute for Plant Protection, Budapest, Hungary

L. M. Schoonhoven (E), Landoew Mogenschool, Wageningen, The Netherlands

Phytochemistry

B. G. Chan, USDA-ARS, Western Regional Research Laboratory, Berkeley, CA 94710

N. H. Fischer, Department of Chemistry, Louisiana State University, Baton Rouge, LA 70803

R. C. Gueldner, USDA-ARS, P.O. Box 5677, Athens, GA 30617

P. A. Hedin, USDA-ARS, Crop Science Research Laboratory, P.O. Box 5367, Mississippi State, MS 39762

R. F. Severson, USDA-ARS, Box 5677, Athens, GA 30613

A. C. Waiss, Jr., USDA-ARS, Western Regional Research Laboratory, Berkeley, CA 94710

Potato

J. Alcazar (E), International Potato Center, Apartado 5969, Lima, Peru

R. Chavez (B), Breeding and Genetics Department, International Potato Center, Apartado Postal 5969, Lima, Peru

M. B. Dimock (E), Crop Genetics International, 7170 Standard Drive, Hanover, MD 21706

P. Gregory (B), International Potato Center, Apartado 5969, Lima, Peru

T. M. Masajo (E), International Institute of Tropical Agriculture PMB 5320, Ibadan, Nigeria

J. J. Obrycki (E), Department of Entomology, Iowa State University, Ames, IA 50011

M. Palacios (E), International Potato Center, Apartado 5969, Lima, Peru

E. B. Radcliffe (E), Department of Entomology, University of Minnesota, St. Paul, MN 55108

K. V. Raman (E), International Potato Center, Apatado 5969, Lima, Peru

W. M. Tingey (E), Deqartment of Entomology, Cornell University, Ithaca, NY 14853

Pulse Crops

L. W. Kitch (legumes) (E), Department of Entomology, Purdue University, West Lafayette, IN 47907

L. L. Murdock (legumes) (E), Department of Entomology, Purdue University, West Lafayette, IN 47907

TABLE 5.2 *(Continued)*

D. F. Palmer (legumes) (E), DeKalb Pfizer Genetics, Rt 2 Box 57, Olivia, MN 56277

Rice

P. C. S. Babu (E), Department of Agricultural Entomology, Tamil Nadu Agricultural University, Madurai, India

P. Caballero (C), Department of Cereal Chemistry, IRRI, P.O. Box 933, Manila, The Philippines

N. Chandramohan (E), Department of Agricultural Entomology, Tamil Nadu Agricultural University, Coimbatore, India 641003

S. Chelliah (E), Department of Agricultural Entomology, Tamil Nadu Agricultural University, Coimbatore, India 641003

Bing-huei Chen (E), Taiwan Agricultural Research Institute, 189 Chung-Cheng Rd., Wan-Feng, Wu-feng, Taichung, Taiwan, Republic of China

R. R. Cogburn (E), USDA-ARS, Rt 7 Box 999, Beaumont, TX 17706

M. Gopalan (E), Tamil Nadu Agricultural University, Agricultural College & Research Institute, Madurar, India 625104

K. Gunathilagaraj (E), Department of Agricultural Entomology, Tamil Nadu Agricultural University, Coimbatore 641 003, India

L. Hai-feng (E), Institute of Plant Protection, Jilin Academy of Agricultural Science, Ginghuling, Jilin, People's Republic of China

E. A. Heinrichs (E), Department of Entomology, Louisiana State University, Baton Rouge, LA 70803

J. Hirao (E), Kyushu National Agricultural Experiment Station, Fukuoka-ken, Japan

R. C. Joshi (E), Department of Entomology, International Rice Research Institute, P.O. Box 933, Manila, The Philippines

G. Khush (B), Department of Plant Breeding, International Rice Research Institute, P.O. 933, Manila, The Philippines

P. Kishore (E) Division of Entomology, Indian Agricultural Research Institute, New Delhi 110012, India

S. S. Lateef (E), International Crops Institute for the Semi Arid Tropics, Patancheru 502 324, AP, Hyderbad, Andhra Pradesh India 502234

K. R. Lee (E), Kon Kuk University, Seoul, Republic of Korea

K. Natarajan (E), Department of Agricultural Entomology, Tamil Nadu Agricultural University, Coimbatore, India 641003

E. M. Nowick (G), Louisiana State University Rice Research Station, P.O. 1429, Crowley, LA 70526

G. A. Palanisamy (E), School of Genetics, Tamil Nadu Agricultural University, Coimbatore, India

A. Pantoja (E), Centro International de Agricultara Tropical Rice Program, Apartado Aereo 6713, Cali, Colombia

P. K. Pathak (E), Department of Entomology, G. B. Pant University of Agricultural & Technology, Pantnagar 263 145 (UP), India

C. Pineada (B), Centro Experimental del Algodon, Managua, D. N., Nicarauga

K. Ramaraju (E), Department of Agricultural Entomology, Center for Plant Protection, Tamil Nadu Agricultural University, Coimbatore, India 641003

J. F. Robinson (E), USDA-ARS, Louisiana State University, Rice Research Station, P.O. Box 1429, Crowley, LA 70526

Z. Rong (E), Institute of Plant Protection, Jilin Academy of Agriculture, Jilin, Gohgzhuli, People's Republic of China

R. C. Saxena (E), Department of Entomology, International Rice Research Institute, P.O. Box 933, Manila, The Philippines

K. Sogawa (E), Hokuriku National Agricultural Experiment Station, Hokuriku, Japan

V. Thirakhupt (E), 30 Sukumvit 36 (Napasub), Bangkok, Thailand 10110

R. Velusamy (E), Department of Agricultural Entomology, Tamil Nadu University, Coimbatore, India 641003

M. S. Venugopal (E), Department of Entomology, Agricultural College, Research Institute, Madurai, India 625104

S. K. Verma (E), Department of Entomology, G. B. Pant University of Agriculture & Technology, Pantnagar 263 145 (UP), India

Q. Wei-Jiu (E), Institute of Plant Protection, Jilin Academy of Agriculture, Jilin, Gohgzhuli, People's Republic of China

W. Yun-sheng (E), Institute of Plant Protection, Jilin Academy of Agriculture, Jilin, Gongzuli, People's Republic of China

Small Grains

L. B. Almaraz (E), Catedra de Cerealicultura, Facultad de Agronomia, Universidad Nacional de la Plata, Casilla de Correo, No 31, 1900 La Plata, Argentina

H. Arridge (E), Tropical Development & Research Institute, College House, Wright's Lane, London, United Kingdom W85SJ

P. J. Bramel-Cox (B), Department of Agronomy, Kansas State University, Manhattan, KS 66506

R. L. Burton (E), USDA-ARS, Plant Science Research Laboratory, 1301 N. Western St., Stillwater, OK 74075

A. G. Cook (E), Tropical Development & Research Institute, College House, Wright's Lane, London, United Kingdom W85SJ

R. G. Dahms (E), 6 Teresa Drive, Bella Vista, AR 72714

TABLE 5.2 *(Continued)*

A. G. O. Dixon (B), Department of Agronomy, Kansas State University, Manhattan, KS 66506

D. L. Dreyer (C), USDA-ARS, Western Regional Research Laboratory, 800 Buchanan St. Berkeley, CA 94710

D. E. Foard (E), Department of Entomology, Purdue University, West Lafayette, IN 47907

J. E. Foster (E), Department of Entomology, Purdue University, West Lafayette, IN 47907

R. L. Gallun (E), Department of Entomology, Purdue University, West Lafayette, IN 47907

D. W. Hagstrum (E), USDA-ARS, Department of Entomology, Kansas State University, Manhattan, KS 66506

T. L. Harvey (E), Kansas State University, Fort Hays Branch Experiment Station, Hays, KS 67601

J. L. Hatchett (E), USDA-ARS, Department of Entomology, Kansas State University, Manhattan, KS 66506

R. P. Hoxie (E), Department of Entomology, Michigan State University, East Lansing, MI 48824

F. Kimmins (E), Tropical Development & Research Institute, College House, Wright's Lane, London, United Kingdom W85SJ

H. J. B. Lowe (E), 3 Bowers Croft, Cambridge CB1 4RP, England

J. A. Lowe (E), International Institute of Tropical Agriculture Cereal Improvement Program, PMB 5320, Ibadan, Nigeria

O. G. Merkle (B), USDA-ARS, Department of Agronomy, Oklahoma State University, Stillwater, OK 74074

H. W. Ohm (B), Department of Agronomy, Purdue University, West Lafayette, IN 47907

D. E. Padgham (E), Tropical Development & Research, Institute, College House, Wright's Lane, London, United Kingdom W85SJ

F. L. Patterson (B), Department of Agronomy, Purdue University, West Lafayette, IN 47907

D. C. Peters (E), Department of Entomology, Oklahoma State University, Stillwater, OK 74074

J. J. Roberts (B), USDA-ARS, Department of Agronomy, Georgia Agricultural Experiment Station, Experiment, GA 30212

I. D. Shapiro (E), Plant Protection Institute, Leningrad, USSR

R. H. Shukle (E), USDA-ARS, Department of Entomology, Purdue University, West Lafayette, IN 47907

V. Smelygenetis (E), Plant Protection Institute, Kiev, USSR

N. K. Vilkova (E), Plant Protection Institute, Leningrad, USSR

TABLE 5.2 *(Continued)*

J. A. Webster (E), USDA-ARS, Plant Science Research Laboratory, 1301 N. Western St., Stillwater, OK 74075

D. E. Weibel (B), Department of Agronomy, Oklahoma State University, Stillwater, OK 74074

S. G. Wellso (E), USDA-ARS, Department of Entomology, Purdue University, West Lafayette, IN 47907

G. Wilde (E), Department of Entomology, Kansas State University, Manhattan, KS 66502

Sorghum

T. L. Archer (B), Texas A & M University Research & Extension Center, Rt 3, Lubbock, TX 79401

V. H. Argandona (E), Facultad de Ciencias Basicas de Farmaceuticas, University of Chile, Casilla 653, Santiago, Chile

H. O. Arriga (E), Catedra de Cerealicultura, Facultad de Agronomia, Universidad Nacional de la Plata, Casilla de Correo 31, La Plata, Argentina 1900

B. M. Bellone (E), Catedra de Cerealicultura, Facultad de Agronomia, Universidad Nacional de la Plata, Casilla de Correo 31, La Plata, Argentina 1900

V. R. Bhagwat (E), ICRISAT Patancheru 502 324, Andhra Pradesh 502234, India

B. C. Campbell (E), USDA-ARS, Western Regional Research Laboratory, Berkeley, CA 94710

H. O. Chidichimo (E), Catedra de Cerealicultura, Facultad de Agronomia, Universidad Nacional de la Plata, Casilla de Correo, 31, La Plata, Argentina 1900

C. Inayatullah (E), Department of Entomology, Oklahoma State University, Stillwater, OK 74074

K. C. Jones (E), USDA-ARS, Western Regional Research Laboratory, 800 Buchanan St. Berkeley, CA 94710

R. J. Molyneux (E), USDA-ARS, Western Regional Research Laboratory, 800 Buchanan St., Berkeley, CA 94710

G. C. Peterson (E), Texas Agricultural Experiment Station, Texas A & M University, Lubbock, TX 79401

W. Reed (E), International Crops Institute for the Semi Arid Tropics, Patancheru 502 324, AP, Hyderbad, Andhra Pradesh India 502234

J. C. Reese (E), Department of Entomology, Kansas State University, Manhattan, KS 66506

C. P. Rumi (E), Catedra de Cerealicultura, Facultad de Agronomia, Universidad Nacional de la Plata, Casilla de Correo, 31, La Plata, Argentina 1900

K. J. Starks, (E), USDA-ARS, Plant Science Research Laboratory, 1301 N. Western St. Stillwater, OK 74075

G. L. Teetes (E), Department of Entomology, Texas A & M University, College Station, TX 77843

TABLE 5.2 *(Continued)*

S. Woodhead (E), Tropical Development & Research Institute, College House, Wright's Lane, London, United Kingdom W85SJ

Soybean

J. M. All (B), University of Georgia, College Experiment Station, Athens, GA 30602

M. Beach (E), Crop Genetics International, 7170 Standard Drive, Hanover, MD 21706

G. L. Beland (E), Funk Seed International, P.O. Box 2911, Bloomington, IL 61701

H. R. Boerma (B), College Experiment Station, Athens, GA

D. J. Boethel (E), Department of Entomology, Louisiana State University, Baton Rouge, LA 70803

T. C. Elden (E), Field Crops Laboratory, USDA-ARS, BARC-E Bldg. 467, Beltsville, MD, 20705

J. H. Elgin, Jr. (B), Field Crops Laboratory, USDA-ARS, BARC-E Bldg. 467, Beltsville, MD 20705

R. B. Hammond (E), Department of Entomology, Ohio Agricultural Research & Development Center, Ohio State University, Wooster, OH 44691

E. E. Hartwig (B), USDA-ARS, Soybean Production Research, P.O. Box 225, Stoneville, MS 38776

J. W. Johnson (B), Department of Agronomy, Georgia Agricultural Experiment Station, Experiment, GA 30212

T. C. Kilen (B), USDA-ARS, Soybean Production Research, P.O. Box 225, Stoneville, MS 38776

M. Kogan (E), Illinois Natural History Survey, 607 East Peabody #172, Champaign, IL 61820

L. L. Lambert (E), USDA-ARS, Soybean Production Research, P.O. Box 225, Stoneville, MS 38776

A. L. Lourencao (E), Institute of Agronomy, 13100 Campinas, Sao Paulo, Brazil

D. M. Norris (E), 237 Russell Laboratories, University of Wisconsin, Madison, WI 53706

G. K. Ruefner (E), Ohio Agricultural Research & Development Center, Ohio State University, 2021 Coffey Road, Wooster, OH 43210

J. W. Todd (E), Department of Entomology, Coastal Plain Experiment Station, Tifton, GA 31794

Sugarcane

O. Sosa, Jr. (E), USDA-ARS, Sugarcane Field Station, Star Route Box 8, Canal Point, FL 33438

W. H. White (E), USDA-ARS, Sugarcane Field Laboratory, P.O. Box 470, Houma, LA 70361

TABLE 5.2 *(Continued)*

Sunflower

G. J. Brewer (E), Department of Entomology, North Dakota State University, Fargo, ND 58105

J. F. Miller (B), USDA-ARS, Department of Agronomy, North Dakota State University, Fargo, ND 58105

Tobacco

J. M. Jackson (E), USDA-ARS, Tobacco Research Laboratory, P.O. Box 1555, Oxford, NC 27565

V. A. Sisson (B), USDA-ARS, Tobacco Research Laboratory, P.O. Box 1555, Oxford, NC 27565

Vegetables

S. Benepal (beans) (E), Virginia State University, P.O. Box 61, Petersburg, VA 23803

M. J. Berlinger (tomato) (E), ARO Gilat Regional Experiment Station, Mobile Post, Negev, Israel 85-280

W. V. Campbell (peanuts) (E), Department of Entomology, P.O. Box 7613, North Carolina State University, Raleigh, NC 27695

W. W. Cantelo (E), USDA-ARS, BARC-E Bldg, Beltsville, MD 27075

A. M. Castro (B), Institute of Vegetable Physiology, Agronomy Faculty, UNLP, Casilla de Correo 31, La Plata, Argentina 1900

R. B. Chalfant (E), Department of Entomology, P.O. Box 748, Coastal Plains Experiment Station, Tifton, GA 31793

R. Dahan (E), Agricultural Research Organization Entomology Laboratory, Moble Post Negev 2, Gilat, Israel 85–200

O. M. B. de Ponti (B), Institute of Horticulture & Plant Breeding, P.O. Box 16,6700 AA, Wageningen, The Netherlands

F. L. Diieleman (E), Department of Entomology, Agricultural University, Binnenhaven 7, Wageningen, The Netherlands

S. S. Duffey (E) (tomatoes), Department of Entomology, University of California, Davis, CA 95616

C. J. Eckenrode (E), Department of Entomology, New York State Agricultural Experiment Station, Geneva, NY 14456

A. H. Eenink (E), Institute of Horticultural Plant Breeding, P.O. Box 16 4600 AA, Wageningen, The Netherlands

P. R. Ellis (E), Department of Entomology, National Vegetable Research Station, Wellesbourne, Warick CV 35 9EF, United Kingdom

K. D. Elsey (E), USDA-ARS, Vegetable Laboratory, 2875 Savannah Highway, Charleston, SC 29407

TABLE 5.2 *(Continued)*

R. L. Fery (E), USDA-ARS, Vegetable Laboratory, 2875 Savannah Highway, Charleston, SC 29407

L. K. French (legumes) (E), French Agricultural Research, Rt 2, Lamberton, MN 56152

J. D. Hansen (E), USDA-ARS, Tropical Fruit & Vegetable Research Laboratory, P.O. Box 4459, Hilo, HI 96720

G. G. Kennedy (E), Department of Entomology, North Carolina State University, Raleigh, NC 27650

A. N. Kishaba (E), USDA-ARS, Texas Agricultural Experiment Station, 2415 E. Highway 83, Weslaco, TX 78596

R. K. Lindquist (tomato) (E), Department of Entomology, Ohio Agricultural Research & Development Center, Ohio State University, Wooster, OH 44691

B. T. McClean (E), Asian Vegetable Research & Development Center, Box 42 Shanhua, Tainan, Tawain Republic of China

J. D. McCreight (E), USDA Boyden Entomology Laboratory, University of California, Riverside, CA 92521

J. P. Moreau (beans) (E), Station Centrale de Zoologie, Agricole C. N. R. A., Versailles, France

C. E. Peterson (onion) (B), Department of Horticulture, University of Wisconsin, Madison, WI 53706

M. F. Ryan (carrot) (E), Department of Zoology, University College Dublin, Belfield, Dublin, 4, Ireland

J. M. Schalk (E), USDA-ARS, Vegetable Laboratory, 2875 Savannah Highway, Charleston, SC 29407

N. S. Talekar (E), Asian Vegetable Research & Development Center, Shanhua, Tainan 741, Taiwan Republic of China

G. A. Wheatley (E), National Vegetable Research Station, Wessesbourne, Warick, United Kingdom

Wheat

A. Barbulescu (E), Research Institute for Cereals & Industrial Crops, 8264 Fundulea, Calarasi, Romania

R. M. DePauw (E), Canadian Agricultural Research Station, P.O. Box 1030, Swift Current, Saskatchewan, S9H 3X2, Canada

A. N. Frolov (E), Plant Protection Institute, Leningrad, USSR

H. N. Lafever (E), Department of Entomology, Ohio Agricultural Research & Development Center, Ohio State University, Wooster, OH 44691

B. Lesczynski (E), Department of Biochemistry, Agriculture & Education College, UL. Prusa (WSR-P) 08-110, Siedlce, Poland

H. M. Niemeyer (E), Canadian Agricultural Research Station, Swift Current, Saskatchewan, Canada

TABLE 5.2 *(Continued)*

S. Niraz (E), Department of Biochemistry, Agricultural & Education College, UL. Prusa (WSR-P) 08-110, Siedlce, Poland

P. L. Raymer (B), Department of Agronomy, Georgia Agricultural Experiment Station, Experiment, GA 30212

A. Y. Semionova (E), Plant Protection Institute, Leningrad, USSR

S. Stamenkovic (E), University of NoviSad, Makisma Gorkog 30, 21000 NoviSad, Yugoslavia

Key to codes: B = plant breeder; E = entomologist; G = geneticist.

TABLE 5.3 World Sources of Crop Germplasm for Evaluation of Plant Resistance to Insects

CASSAVA AND BEANS

International Center for Tropical Agriculture (CIAT), Palmira, Colombia

COWPEA AND SWEET POTATO

International Institute of Tropical Agriculture (IITA), Ibadan, Nigeria

MAIZE

Instituto Colombiana Agropecuorio, Medellin, Colombia

Instituto Nacional de Investigaciones Agricolas (INIA), Chapingo, Mexico

International Maize and Wheat Improvement Center (CIMMYT), El Batan, Mexico

International Center for Insect Physiology and Ecology (ICIPE), Nairobi, Kenya

International Crops Research Institute for the Semi-Arid Tropics (ICRISAT), Hyderabad, India

Pioneer Hi-Bred International, Inc., Union City, TN

Northrup, King & Co., Eden Prairie, MN

USDA-ARS Maize Germplasm Collection, Ames, IA

POTATO

International Potato Center (CIP), Lima, Peru

USDA-ARS Interregional Potato Introduction Station, Sturgeon Bay, WI

RICE

National Agricultural Research Institute, Non Repos, East Coast Demerara, Guyana

International Rice Research Institute (IRRI), Manila, Philippines

USDA National Rice Collection, Aberdeen, ID

TABLE 5.3 *(Continued)*

SMALL GRAINS (BARLEY, OATS, RYE)

Waite Agricultural Research Institute Barley Collection, Adelaide, Australia
Ohara Institute for Agricultural Biology, Okayama University, Kurashiki, Japan
Research Institute of Cereals, Havlivkovo, Czechoslovakia
Swedish Seed Association, Svalov, Sweden
USDA-ARS National Small Grains Collection, Aberdeen, ID

SORGHUM

International Crops Research Institute for the Semi-Arid Tropics (ICRISAT), Hyderabad, India
Northrup King & Co., Eden Prairie, MN
Texas A & M Sorghum Germplasm Collection, Mayaguez, Puerto Rico

SOYBEAN

Asian Vegetable Research and Development Center (AVRDC), Tainan, Taiwan
USDA-ARS National Soybean Collection, USDA Bioenvironmental Laboratory, Stonville, MS
Soybean Germplasm Collection, Department of Agronomy, University of Illinois Champaign, IL

SUGARCANE

Indian National Sugarcane Germplasm Collection, Coimbatore, India
USDA-ARS National Sugarcane Germplasm Collection, Canal Point, FL

WHEAT

Australian Wheat Collection, Tamworth, New South Wales, Australia Plant Breeding Institute, Cambridge, England
Laboratorio del Germoplasmo, Bari, Italy
Crop Research and Introduction Center, Ismir, Turkey
DeKalb Agricultural Research, Inc., Mason, MI
USDA National Wheat Collection, Aberdeen, ID
North American Plant Breeders, Brookston, IN
Northrup King & Co., Eden Prairie, MN

MULTISPECIES COLLECTIONS

National Institute of Agricultural Science, Hiratsuka, Japan
USDA-ARS National Seed Storage Laboratory, Fort Collins, CO
USDA-ARS Northeastern Regional Plant Introduction Station, Geneva NY
USDAP-ARS Southern Regional Plant Introduction Station, Experiment, GA
USDA-ARS North Central Regional Plant Introduction Station, Ames, IA
USDA-ARS Western Regional Plant Introduction Station, Pullman, WA

plant material in the United States is available to agricultural scientists for experimental use. In order to maintain collections, however, borrowers are frequently requested to return samples of seed produced by plants in experimental plantings. Many of the collections are very large and few have been thoroughly evaluated under controlled conditions for insect resistance to several pests.

In addition to the agencies listed above that distribute plant material, agencies exist to provide advice to governmental officials about the status of plant material. In the United States, the National Plant Genetic Resources Board exists to advise the Secretary of Agriculture and National Association of State Universities and Land Grant Colleges about national plant material needs. This agency also recommends plans to coordinate the collection, maintainence, description, evaluation, and ultilization of plant material between the United States and international organizations.

5.3 CONDITION OF EXISTING PLANT MATERIAL STOCKS

Efforts by the USDA National Plant Material System and several International Research Centers are currently in progess to collect, preserve, and maintain plant material of the major food crops of the world with as much genetic diversity as possible. These efforts are necessary to avoid the occurrence of outbreaks of diseases and insects among crop plants with similar cytoplasmic factors for susceptibility (Horlan 1972). Such an event occurred in 1970 in the United States when a southern corn leaf blight epidemic affected the majority of the corn hectarage in the southern United States. Justifiably, governmental agencies should provide funding to improve the holdings of national plant material collections and remove the "genetic vulnerability" from national seed storage collections (Norman 1981; Walsh 1981a, 1981b).

There are currently major shortages of genetic diversity in cassava, maize, millet, potato, and sorghum plant material stocks (Table 5.4). There is also concern by some members of the International Board of Plant Germplasm Resources that rice and wheat plant material collections may be especially inadequate in their content of wild species (Hargrove et al. 1980, 1985).

Global plant material preservation efforts are currently in serious jeopardy however, owing to slash and burn agricultural practices, population expansion, and unchecked timber and mining activities in much of the humid tropics of the world (Raven 1983). Hundreds of species of plants with potential medical and agricultural uses are slowly disappearing in the tropical regions of the world.

TABLE 5.4 Present Size and Need for Improvement of the Major World Food Crop Germplasm Collections

Crop	Approximate No. of Accessions	Need for Improvements
Wheat	35,000	Small
Maize	20,000	Urgent
Rice	35,000	Small
Barley	18,000	Moderate
Potato	3,000	Urgent
Soybean	4,000	Moderate
Sugarcane	1,000	Small
Sorghum	17,000	Critical
Millet	1,000?	Critical
Cassava	1,000?	Urgent

With decreasing amounts of wild plant material available for use in many crop plant species, it is more necessary than ever to preserve existing global crop plant material collections better. Additional efforts are now necessary to increase the diversity and amount of collections and to make efforts to collect new genetic materials that can be incorporated into domestic crop plant species and further broaden the genetic composition of these species. Activity by plant resistance researchers in both areas is expressly needed. There are many opportunities available for close interdisciplinary research between entomologists and plant breeders to conduct these studies.

REFERENCES

Hargrove, T. R., W. R. Coffman, and V. L. Cabanilla. 1980. Ancestry of improved cultivars of Asian rice. Crop Sci. 20:721–727.

Hargrove, T. R., V. L. Cabanilla, and W. R. Coffman. 1985. Changes in rice breeding in 10 Asian countries: 1965–1984. Diffusion of genetic materials, breeding objectives, and cytoplasm. IRRI Research Paper Series No. 111. 18 pp.

Harlan, J. R. 1972. Genetics of disaster. J. Environ. Qual. 1:212–215.

Harris, M. K. 1975. Allopatric resistance: Searching for sources of insect resistance for use in agriculture. Environ. Entomol. 4:661–669.

Heinrichs, E. A. 1986. Perspectives and directions for the continued development of insect-resistant rice varieties. Agric. Ecosyst. Environ. 18:9–36.

Jennings, P. R. and A. Pineda. 1970. Effect of resistant rice plants on multiplication of the plant hopper, *Sogatodes orizicola* (Muir.). Crop Sci. 10:689–690.

Leppick, E. E. 1970. Gene centers of plants as sources of disease resistance. Ann. Rev. Phytopathol. 8:324–344.

Norman, C. 1981. The threat to one million species. Science 214:1105–1107.

Panda, N. and E. A. Heinrichs. 1983. Levels of tolerance and antibiosis in rice varieties having moderate resistance to the brown planthopper, *Nilaparvata lugens* (Stal) (Hemiptera:Delphacidae). Environ. Entomol. 12:1204–1214.

Pathak, M. D. 1977. Defense of the rice crop against insect pests. Ann. N.Y. Acad. Sci. 287:287–295.

Raven, P. 1983. The challenge of tropical biology. Bull. Entomol. Soc. Am. 29:5–12.

Walsh, J. 1981a. Genetic vulnerability down on the farm. Science 214:161–164.

Walsh, J. 1981b. Plant material resources are losing ground. Science 214:431–432.

6 Techniques Used to Measure Plant Resistance to Insects

6.1 MANIPULATION OF INSECT POPULATIONS

6.1.1 Field Populations

In order to determine if insect resistance exists in a diverse group of plant material, it is necessary to manipulate either the pest insect population or the test plant population. Rarely is a researcher able simply to plant a group of plant material and accurately evaluate the insect damage it sustains. Without proper planning, the researcher will have either plants with insufficient insect populations to inflict damage or insect populations available to infest plants that are at an improper phenological stage of development.

Field populations of pest insects are normally used by researchers to evaluate plant materials in the early stages of a plant resistance program. Obviously, the final proving ground for plant material found to be resistant is in replicated field tests. However, the field evaluation of plant material has some inherent problems that may directly affect the search for resistance. Unmanaged insect populations may be too low or unevenly distributed in space and/or time to inflict a consistent level of damage. Year to year variation in population levels of the target pest insect may also make interpreting the results of field evaluations difficult. Finally, unmanaged field populations may be contaminated with nontarget pest insects that have feeding symptoms similar to those of the target pest.

Managed or supplemented populations will ensure a more uniform distribution of insects, but these insects are subject to mortality by naturally occurring biological control agents such as predators, parasites, or pathogens. For this reason, test plants should be treated with a selective insecticide before infestation to eliminate populations of beneficial insects. Insects on test plants should be subsequently protected from predators by caging or by repeated applications of beneficial-specific insecticides. In spite of these precautions, some supplemented insect populations may suffer high mortality from pathogen infection, owing to the abnormally high insect population densities occurring in cages. The most useful insect population is the one that causes sufficient damage for the researcher to observe the maximum differences among plant material evaluated.

Several procedures can be employed to obtain a useful field population of test insects. Begin by planting a trap crop consisting of a mixture of susceptible cultivars of differing maturities. This mixture should be planted as border rows around the actual test plots and at regular intervals within the plots. Ideally, the trap crop mixture should occur after every experimental row, but this may not be possible, because of total field plot space limitations. If large-scale populations of pest insects fail to build up in the test plots, trap crop rows can be mechanically cut to force insects onto test plants. If the supply of experimental plant material is sufficient, the entire experiment may be planted again in duplicate or triplicate so that at least one planting will coincide with the peak pest insect population level in the field.

Selective insecticides can also be applied to plants in field plots to eliminate nontarget pests, predators, or parasites of naturally occurring pest insect populations, causing a resurgence in population levels (Chelliah & Heinrichs 1980). Pest insect populations can also be augmented by placing light traps, pheromone traps, or kairomone traps to draw insects into the experimental plots. Finally, researchers may find it useful to make mass collections of indigenous pest insect populations collected from surrounding areas and re-release them onto test plants.

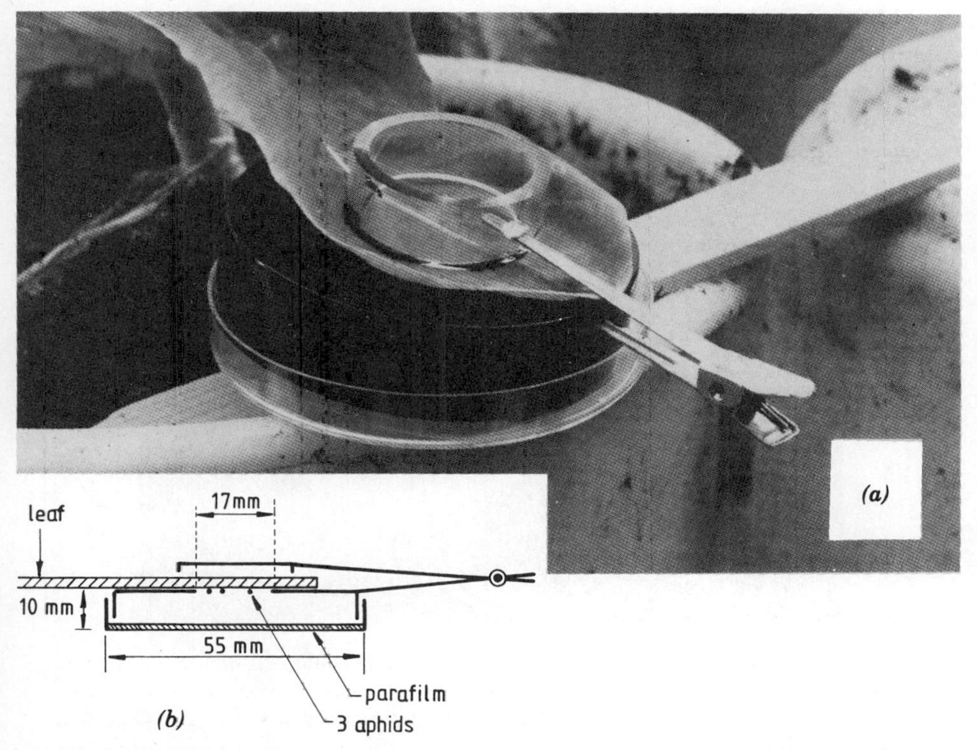

FIGURE 6.1 (*a*) Cage for collecting honeydew produced by aphids feeding on lettuce cultivars. (*b*) Cross section of cage. (From Eenink et al. 1984. Reprinted with the permission of Nijhoff Publishing Co.)

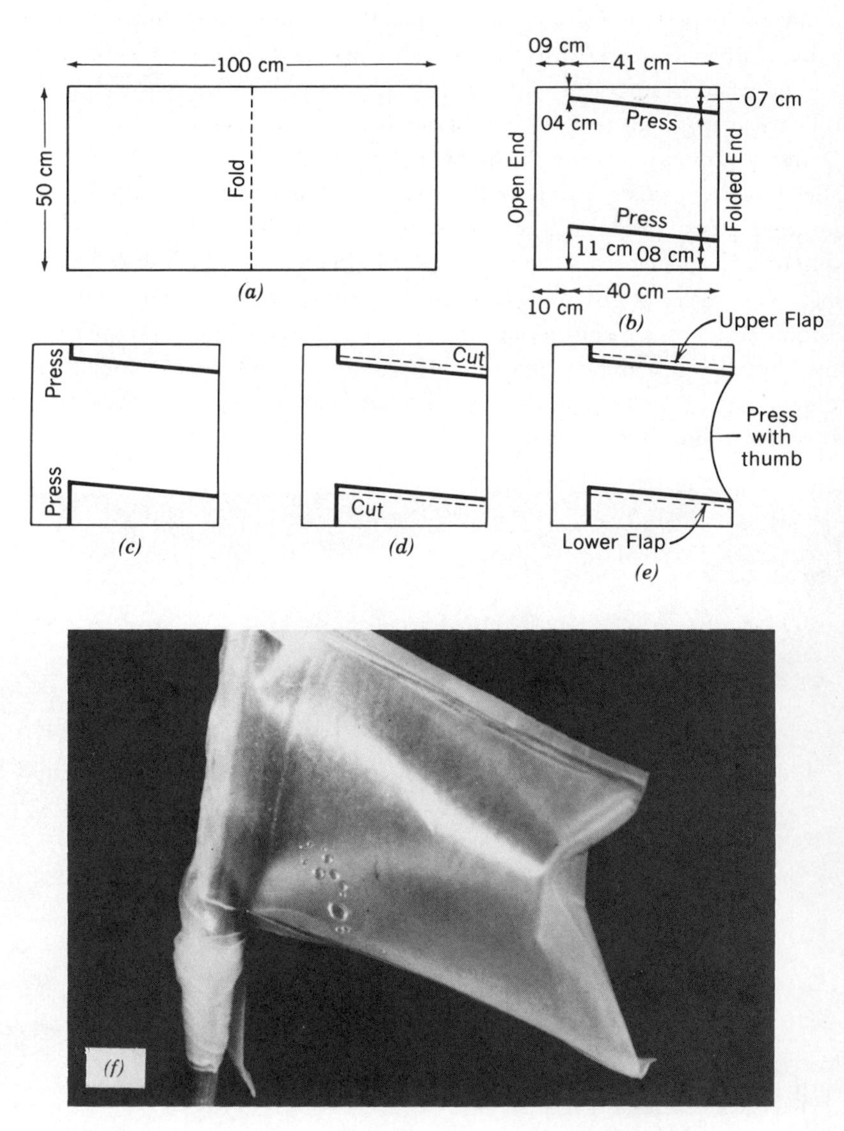

FIGURE 6.2 (*a–e*) Steps involved to prepare a Parafilm sachet for collecting honeydew excreted by the brown plant hopper, *Nilaparvata lugens Stal*, while feeding on rice stems. (*f*) Completed sachet showing honeydew collected fron one female feeding for 24 hours. (From Pathak et al. 1982. Reprinted with permission from J. Econ. Entomol. 75:194–195. Copyright 1982, Entomological Society of America.)

6.1.2 Caged Insect Populations

In spite of all of the efforts outlined above, caging insects on test plants may be necessary. Cages offer two major advantages to the plant resistance researcher. They limit emigration of the test insect from plants being evaluated, and they protect these insects from predation and parasitism. In greenhouse and field tests, insects may be housed in small clip-on cages over small sections of intact plant leaves, stems, or flowers (Fig. 6.1). Pathak et al. (1982) developed a Parafilm sachet cage to collect honeydew from leafhoppers and plant hoppers feeding on resistant and susceptible rice cultivars (Fig. 6.2). A modification of these cages has been developed by Tedders and Wood (1987). Larger plant tissue can be enclosed in sleeve cages made of dialysis tubing (Fig. 6.3), polyester organdy or nylon cloth (Fig. 6.4), or hydroponic plant growth pouches (Fig. 6.5). Special modifications of plant growth conditions, such as the slant board technique, have also proven useful for evaluation of legumes for resistance to root feeding insects (Byers & Kendall 1982, Powell et al. 1983).

Whole plants can be placed in cages constructed of wood, Plexiglas®, or metal frames, supporting screened aluminum panels of nylon or saran (Figs. 6.6 and 6.7). Cage size and shape are determined by the type and number of test plants that must be evaluated. Dimensions vary from small Cornell type cages to large cages that are annually placed over galvanized metal frames to cover entire field experiments (Fig. 6.8) (Lambert 1984). Saran screening, a polyester coated nylon material, offers superior resistance to environmental deterioration. Cage openings can be closed using heavy duty zippers or the artificial

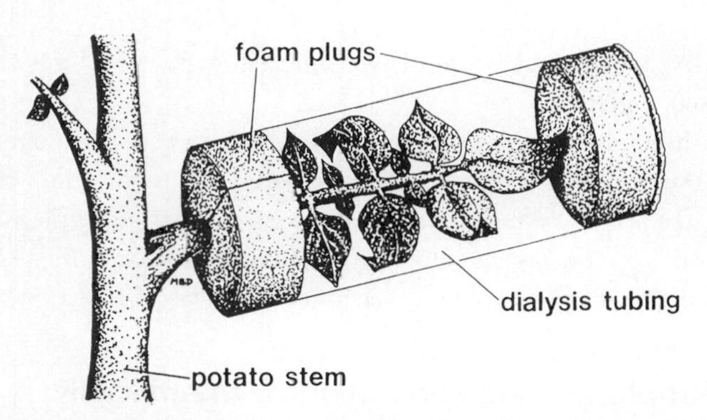

FIGURE 6.3 Dialysis tubing sleeve cage with polyurethane foam end plugs used to confine Colorado potato beetle, *Leptinotarsa decemlineata* (Say), larvae on leaves of *Solanum* spp. plants. (From Dimock et al. 1986. Reprinted with permission from J. Econ. Entomol. 79:1269–1275. Copyright 1986, Entomological Society of America.)

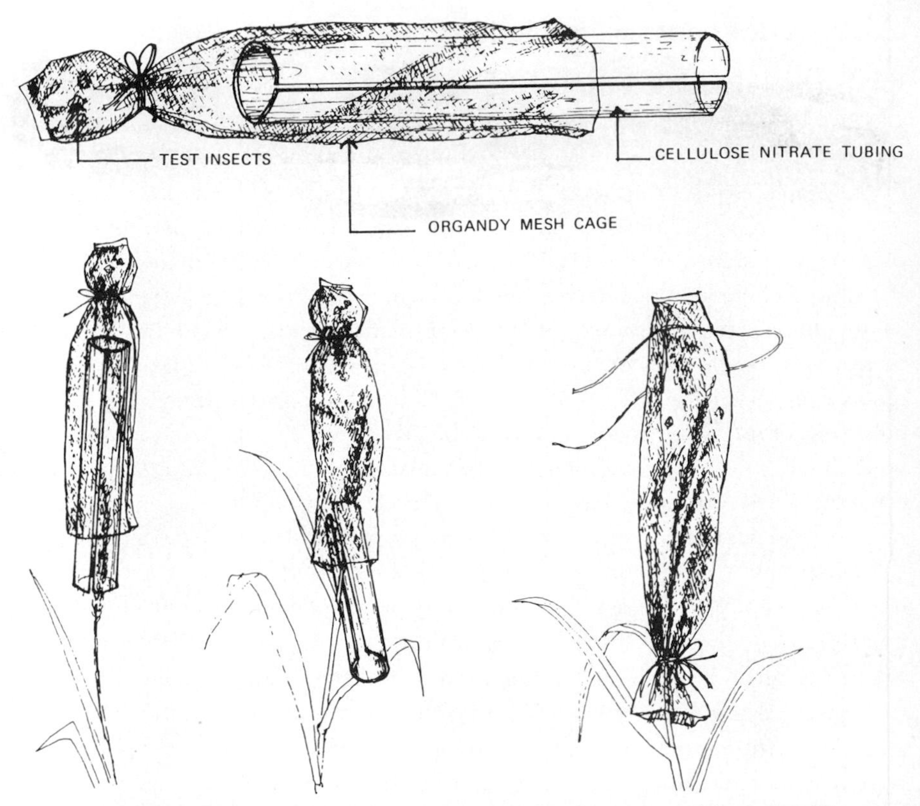

FIGURE 6.4 Organdy mesh sleeve cage used to confine rice stink bug, *Oebalus pugnax* (F.), adults on panicles of rice plants (from Hollay et al. 1987).

fiber Velcro®, which makes use of two different types of fabrics that "lock" when pressed together.

Despite their advantages, however, cages also have some inherent disadvantages that should be anticipated and may require compensation. Some cages may cause abnormal environmental conditions that can alter plant growth or cause foliar disease outbreaks. Obviously, not all plants are affected similarly, and cage affects, if any must be determined on a case-by-case basis.

6.1.3 Supplementing Populations with Artificially Reared Insects

If the pest insect can be mass reared, then insects are available on a year-round basis for evaluating plant material. Scores of artificial diets for rearing insects

FIGURE 6.5 Hydroponic plant growth pouch used to confine larvae of the western corn rootworm, *Diabrotica vigifera vigifera* LeConte, on roots of a maize plant. (From Ortman and Branson 1976. Reprinted with permission from J. Econ. Entomol. 69:380–382. Copyright 1976, Entomological Society of America.)

FIGURE 6.6 Saran screen top cages constructed from the shell of a wooden planting flat to confine adult Mexican bean beetles, *Epilachna varivestis* Mulsant, on seedling soybean plants.

FIGURE 6.7 Screen cages placed over rows of small grain plants to confine cereal leaf beetles, *Oulema melanoplus* (L.). (From Webster and Smith 1983. Reprinted through the courtesy of the U.S. Department of Agriculture.)

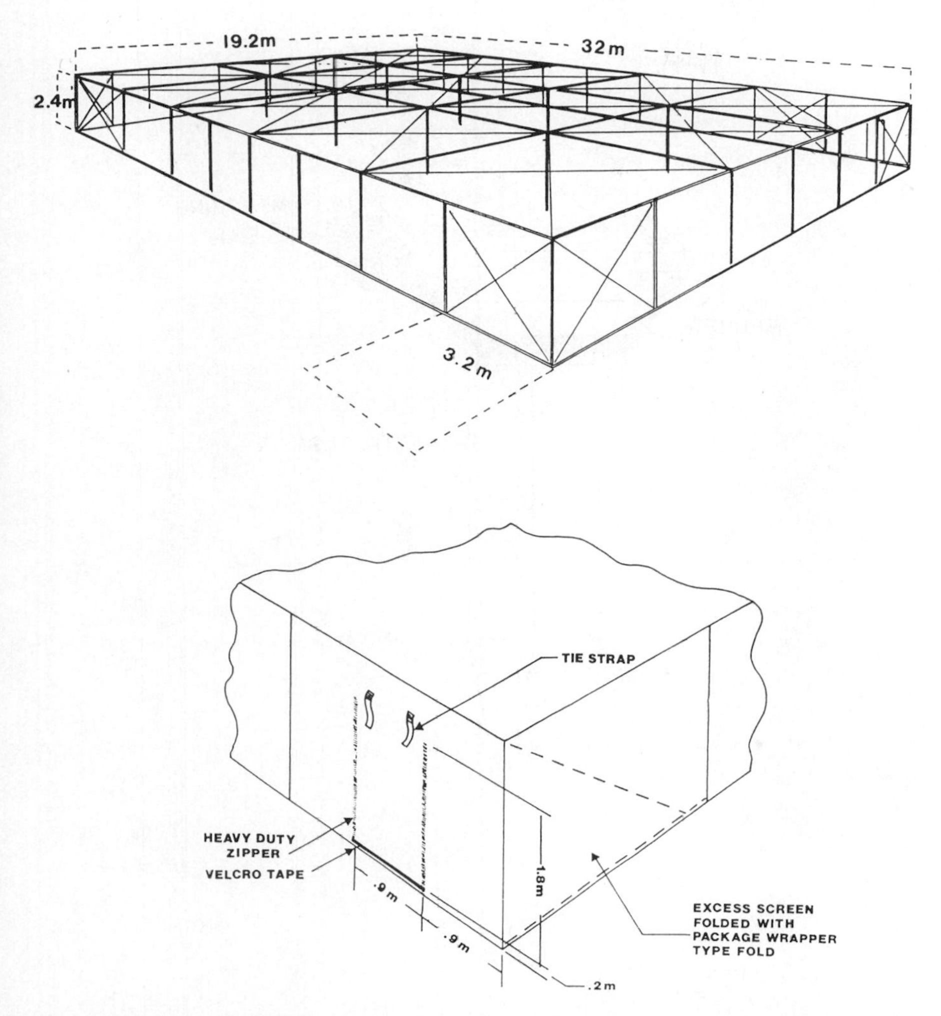

FIGURE 6.8 Galvanized permanent metal cage frame supporting saran screen cover used to confine insect populations on field plantings of soybeans. (From Lambert 1984. Reprinted with the permission of the American Society of Agronomy, Inc.)

have been developed. For complete discussions, readers are referred to Singh and Moore (1985). The greatest successes in insect mass rearing for plant resistance evaluations have been with foliar and stalk feeding Lepidoptera. Commercial maize hybrids with resistance to the European corn borer, *Ostrinia nubilalis*, Hubner, the southewestern corn borer, *Diatrea grandiosella* (Dyar), and the fall armyworm, *Spodoptera frugiperda* (J. E. Smith), have been produced

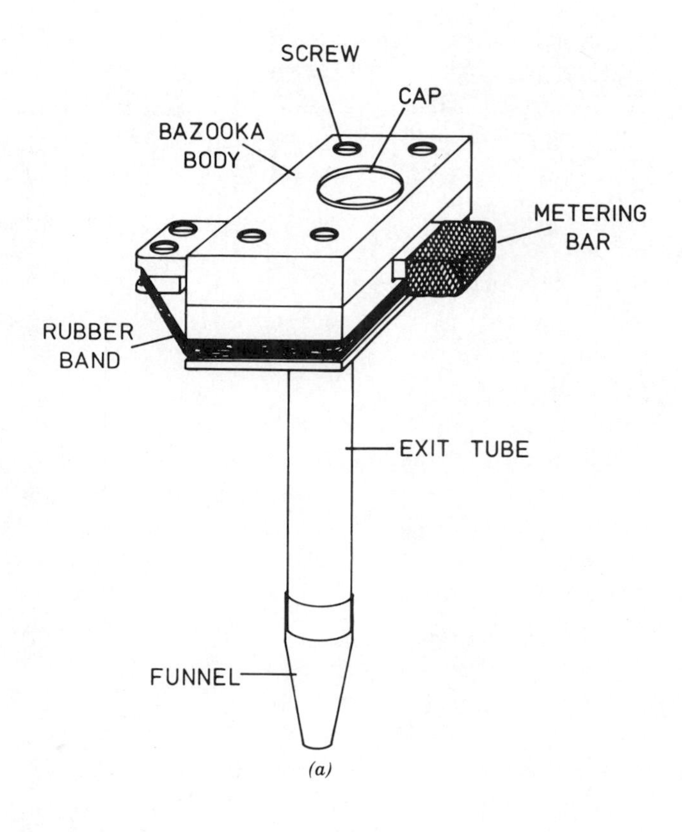

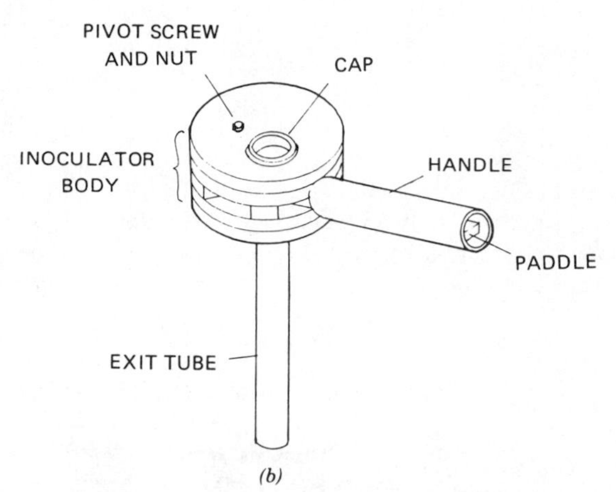

FIGURE 6.9 Manual dispensers for innoculating plant tissues with immature lepidoterous larvae. (*a*) Modified 'Bazooka' inoculator (from Mihm et al. 1978); (*b*) Davis larval inoculator (From Davis & Oswalt 1979.)

because of the development and refinement of techniques to handle and rear these insects (Mihm 1983a–c). Mechanical techniques have also been developed to reduce greatly the amount of time required to mix, dispense, and inoculate artificial diets (Davis 1980a, b), remove pupae from diet (Davis 1982), and harvest insect eggs (Davis 1982, Davis et al. 1985). The net result of these accomplishments has been a quantum increase in the annual amount of plant material that can be accurately evaluated.

Techniques to infest plants have also been developed and refined. Early methods made use of agar-based suspensions containing corn earworm, *Heliothis zea* Boddie, eggs that were injected with maize silk masses (Widstrom & Burton 1970). Similar techniques have been used to apply bollworm, *Heliothis virescens* (F.) eggs to the fruiting structures of cotton plants (Dilday 1983). The latest refinement in plant inoculation technology is a plastic manual *larval plant inoculator* (Fig. 6.9) that dispenses a suspension of immature larvae and fine-mesh sterilized corncob grits onto plant tissues (Wiseman et al. 1980). This inoculation method allows rapid, accurate placement of larvae onto plants (Wiseman & Widstrom 1980, Davis & Williams 1980). The larval inoculator, first termed a "Bazooka" by Mihm et al. (1978), has been used successfully to infest test plants with several species of insects (Table 6.1).

Continued production of test insect populations on artificial diet often decreases their genetic diversity (Berenbaum 1986). To avoid these problems, quality control measures must be made a part of the rearing program to ensure that the behavior and metabolism of the laboratory reared insect is similar to that of wild individuals. An effective means of avoiding the development of these problems is to infuse wild individuals into the laboratory colony and to

TABLE 6.1 Insects Successfully Dispensed onto Crop Plant Cultivars in Plant Resistance Evaluations Using Inoculator

Crop and Insect	References
MAIZE	
Bollworm, *Heliothis virescens* (F.); corn earworm, *Heliothis zea* (Boddie); European corn borer, *Ostrinia nubilalis* (Hubner); fall armyworm, *Spodoptera frugiperda* (J. E. Smith); southwestern corn borer, *Diatraea grandiosella* (Dyar)	Davis 1980b, Mihm 1983a–c
RICE	
Fall armyworm, *Spodoptera frugiperda* (J. M. Smith)	Pantoza et al. 1986

ensure that the artificial diet closely resembles the nutritional and allelochemical composition of the host plant.

6.2 WHEN AND HOW TO EVALUATE TEST PLANTS

6.2.1 Methods Used to Differentiate Between Resistance Categories

Different experimental test procedures are necessary to differentiate between the antixenosis, antibiosis, and tolerance categories of plant resistance to insects. Much of the effort in a program to develop insect-resistant cultivars involves elimination of susceptible plant materials. Therefore, large-scale evaluations where insects are offered a free choice of plant materials, either in field plots or greenhouse experiments, are often conducted initially. Materials identified from these tests as potentially resistant are then reevaluated in a smaller group that includes a susceptible control cultivar. To confirm the antixenosis category of resistance, plant materials are planted and infested together within each experimental replication (field plot, greenhouse flat, or pot). (Different cultivars under evaluation are often planted in a circular arrangement in greenhouse pots and test insects released in the center of the test plants). Test insect populations are left on plants until the susceptible control cultivars have sustained heavy damage or population accumulation and plants are then evaluated for insect feeding damage and/or insect population levels (see Chapter 2). By identifying resistant plants in choice tests, the researcher is assured that the potentially resistant material possesses antixenosis.

To identify antibiosis, plant materials are planted, caged, and infested separately (Davis 1985). Test insects have no choice but to feed or not feed on plants of each cultivar being evaluated. Antibiosis measurements related to insect survival and development such as those discussed in Chapter 3 are then recorded during the course of the development of insects used to infest test plants. Antibiosis and antixenosis are not always easily distinguishable from one another. This is especially evident when experiments are conducted with early stages of immature insects (Horber 1980). These two resistance categories may also be difficult to separate, for the death of test insects in an antibiosis test may result from either the toxic factor(s) involved in antibiosis or the deterrent factor(s) involved in antixenosis (Kishaba & Manglitz 1965, Renwick 1983).

Entirely different techniques are employed to assess plant tolerance, because it does not involve a plant interaction with insect behavior or physiology.

Normally, the existence of tolerance is determined by comparing the production of plant biomass (yield) in insect-infested and noninfested plants of the same cultivar. Yield differences between infested and uninfested plants can then be used to calculate percent yield loss based on the ratio yield of infested plants/yield of uninfested plants. A tolerance evaluation involves preparing replicated plantings that include the different cultivars being evaluated and a susceptible control cultivar, caging all plants in each replicate, and infesting caged plants in one-half of each replicate with insect populations at or above the economic injury level for that insect. Plants should remain infested until susceptible controls exhibit marked growth reduction or until the pest insect has completed at least one generation of development. Volumetric or plant biomass production measurements (see Chapter 4) are then made to calculate percent yield loss measurements.

Standardized evaluation methods exist for the evaluation of the different categories of insect resistance in alfalfa (Nielson 1974, Hill & Newton 1972, Simonet et al. 1978, Sorenson 1974); cassava (Schoonhoven 1974); cotton (Benedict 1983, Dilday 1983, George et al. 1983, Jenkins et al. 1983, Leigh 1983, Schuster 1983, Tugwell 1983); rice (Heinrichs et al. 1985); maize (Guthrie et al. 1960, 1978, Mihm 1983a–c); sorghum (Johnson & Teetes 1979, Soto 1972, Starks & Burton 1977); sugarcane (Agarwal et al. 1971, Martin et al. 1975); and wheat (Webster & Smith 1983). For further reading see the reviews of Davis (1985) and Tingey (1986).

6.2.1.1 Seedlings

Greenhouse experiments allow the researcher to make large-scale evaluations of seedlings in a relatively short period of time. This technique is commonly used to evaluate plant material for resistance to leaf and stem feeding insects, and has proven to be beneficial in eliminating large numbers of susceptible plants (Table 6.2). However, some plant material resistant as seedlings may be susceptible in later growth stages, necessitating field verification of seedling resistance.

6.2.1.2 Mature Plants

If insect damage occurs in the later vegetative stages or in the reproductive stages of plant development, field or greenhouse tests should be conducted with mature plants, regardless of the increased amounts of time, space and labor necessary to grow plants to the age of evaluation. Rufener et al. (1987) described a laboratory-greenhouse procedure that evaluates soybean plants in

TABLE 6.2 Crop Plants Evaluated as Seedlings for Resistance to Insect Pests

Crop Plant and Insect(s)	Reference
ALFALFA	
Potato leafhopper, *Empoasca fabae* (Harris)	Sorensen and Horber 1974
Pea aphid, *Acrythosiphum pisum* (Harris)	Sorensen 1974
Spotted alfalfa aphid, *Therioaphis maculata* (Buckton)	Nielson 1974
Meadow spittle bug, *Philaneous spumaris* (L.)	Hill and Newton 1972
CLOVER	
Clover head weevil, *Hypera meles* (F.)	Smith et al. 1975
COTTON	
Lygus hesperus (Knight)	Leigh 1983
Two-spotted spider mite, *Tetranychus urticae* (Koch)	Schuster 1983
RICE	
Armyworm, *Mythimna separata;* caseworms, *Nymphula depunctalis* (Guenee); gall midge, *Orseolia oryzae* (Wood-Mason); green leafhopper, *Nephotettix virescens* (Distant); green rice hopper, *Nephotettix cincticeps* Uhler; green leafhopper, *Nephotettix nigropictus* (Stal); brown plant hopper, *Nilaparvata lugens* (Stal); white-backed plant hopper, *Sogatella furcifera* (Horvath)	Heinrichs et al. 1985
SORGHUM	
Greenbug, *Schizaphis graminum* (Rondani)	Starks and Burton 1977
WHEAT	
Lepidopterous larvae	Soto 1972
Cereal left beetle, *Oulema melanopus* (L.)	Webster and Smith 1983
Yellow sugarcane aphid, *Sipha flava* (Forbes)	Merkle and Starks 1985

the vegetative stage of development for resistance to feeding by larvae of Mexican bean beetle, *Epilachna varivestis* Mulsant. Identification of resistant plants before flowering allows crosses involving these plants to be made in the same growing season and reduces the time required to develop beetle-resistant cultivars. In field studies, planting dates should be adjusted to coincide with the expected time of peak insect abundance. If necessary, two or three separate plantings should be made over time, in order to have one planting that best coincides with the insect population peak.

6.2.2 Altered Plant Tissues

Once resistance has been positively identified, several methods can be employed to alter the configuration of plant tissues, in order to conduct in-depth determinations of the factors that mediate resistance. A common initial

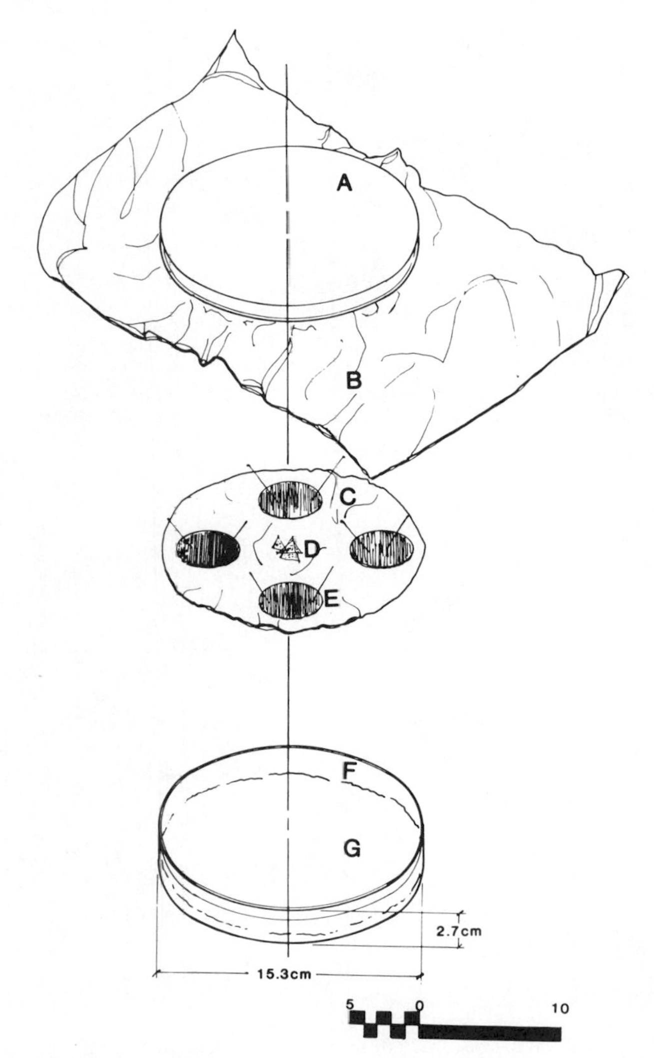

FIGURE 6.10 Polystyrene plastic cylinder chamber for antixenosis tests of excised maize leaf discs. A, Chamber lid, B, sterile tissue, C, Cellucotton® dampened with distilled water, D, lepidopterous egg masses, E, leaf disc, F, chamber, G, 2% agar plus mold inhibitors. (Courtesy F. M. Davis and the U.S. Department of Agriculture.)

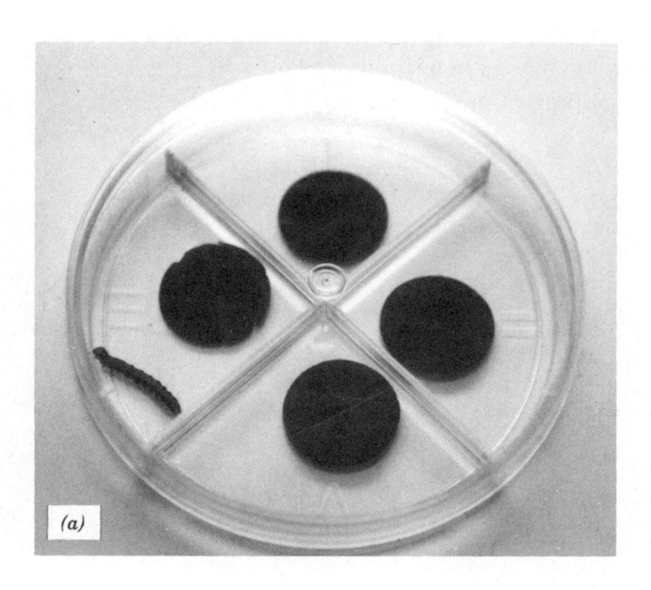

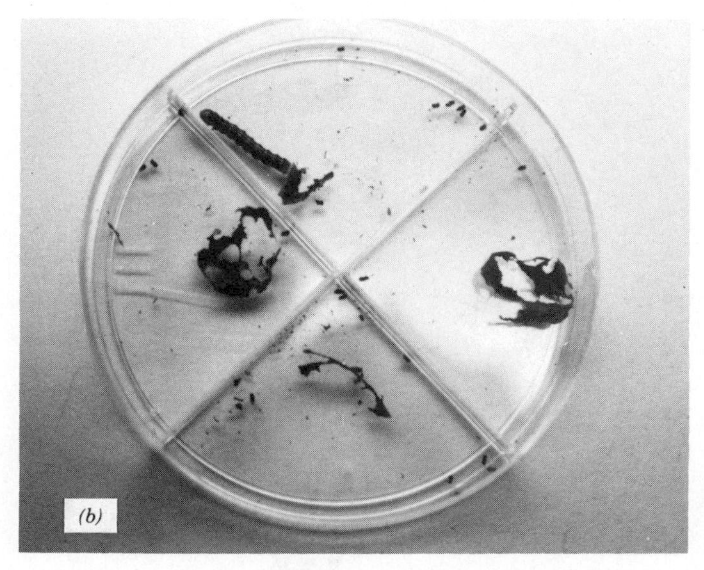

FIGURE 6.11 Plastic petri dish with divided base for evaluating soybean leaf discs for resistance to the corn earworm, *Heliothis zea* (Boddie). Quadrants II & III resistant, quadrants I & IV susceptible. (*a*) Pre-test, (*b*) 24 hours post-test.

procedure is to cut circular discs of plant leaf or root tissue for laboratory evaluation of several genotypes simultaneously (Figs. 6.10 and 6.11). At this stage, the physical structures involved in resistance (i.e., trichomes) can be mechanically removed to determine the effects of these structures on insect behavior (Gibson 1976, Khan et al. 1986).

Tissues can then be altered physically by drying and grinding, followed by removal of extractable phytochemicals by solvent extraction (Fig. 6.12). Extracts can then be assayed for deterrence by applying them onto inert substrates such as discs of filter paper, polyurethane foam (Ascher & Nemny 1978), glass fibers (Adams & Bernays 1978, Stadler & Hanson 1976), and cellulose nitrate membrane filters (Bristow et al. 1979, Doss & Shanks 1986). The concentrations of allelochemicals in this type of assay substrate may not

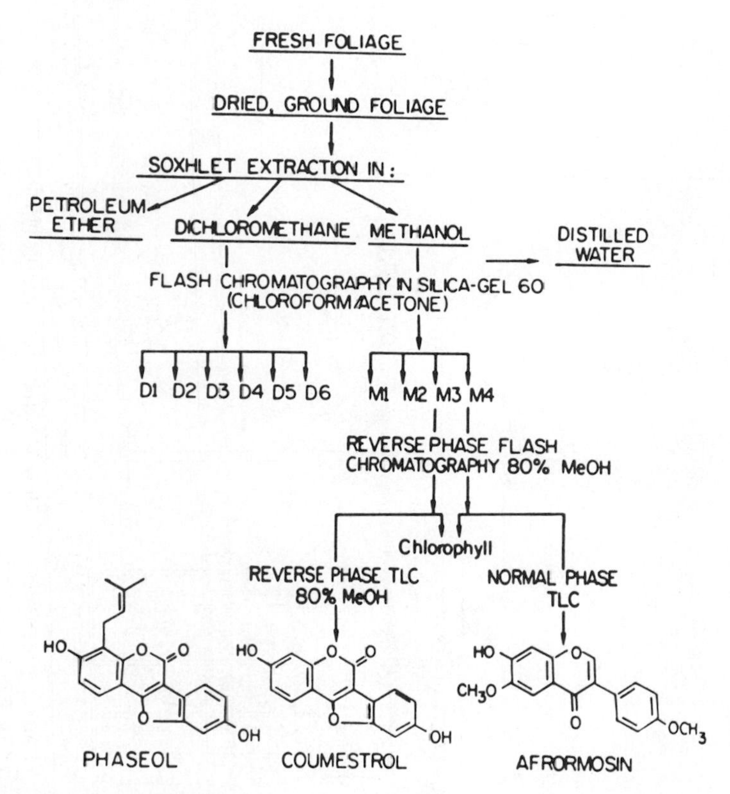

FIGURE 6.12 Extraction procedure used to isolate allelochemicals from soybean for bioassay with the soybean looper, *Pseudoplusia includens* (Walker). (From Caballero 1985. Reprinted with the permission of the author.)

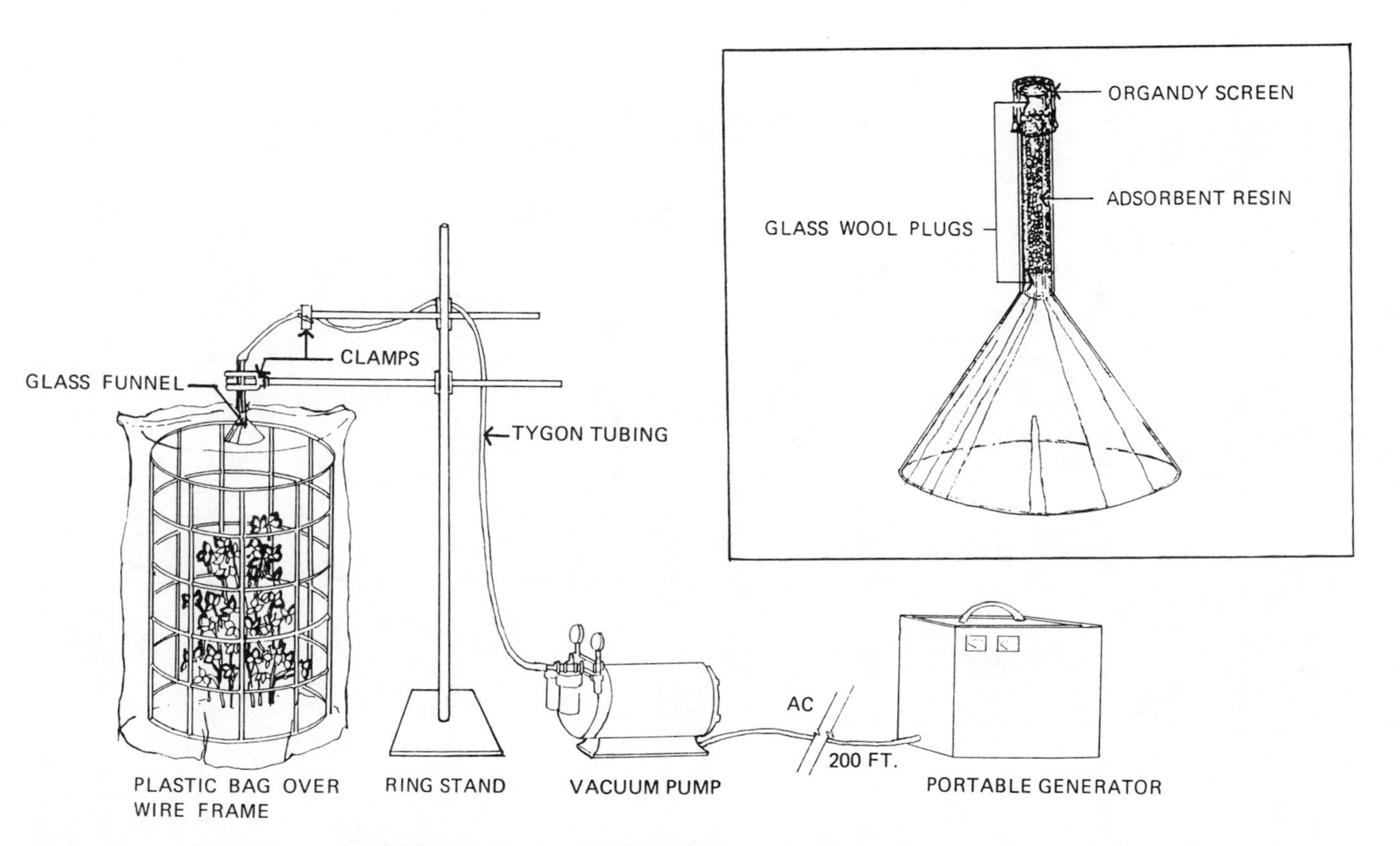

FIGURE 6.13 Apparatus used to collect plant volatiles.

always be the biological equivalent of those occurring in living plant tissues; Woodhead (1983) found that 75–80% of the chemicals applied to glass fiber discs were located along the disc periphery. Extracts can also be evaluated for their effects on insect growth and metabolism by adding them in solution to an inert substrate such as cellulose. After removal of solvent, the extract amended cellulose "cake" is incorporated into the artificial diet mixture for infestation and bioassay (Chan et al. 1978). Unextracted leaf powers (Smith & Fischer 1983, Quisenberry et al. 1988) and homogenized fresh plant parts (Wiseman

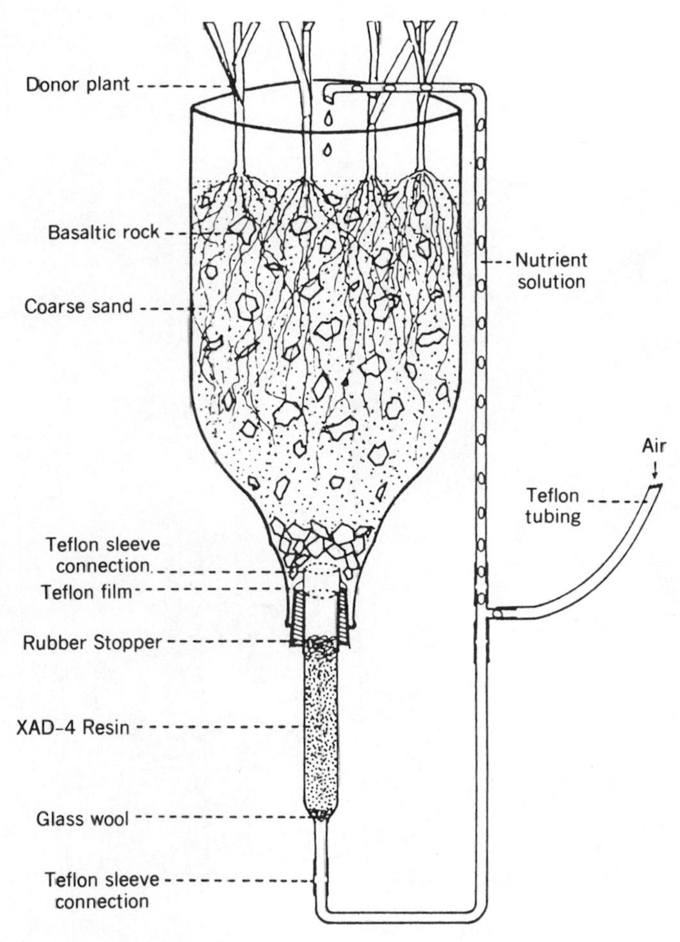

FIGURE 6.14 Hydrophobic plant root exudate trapping system. (From Tang & Young 1982. Reprinted with permission of John Wiley & Sons, Inc.)

et al. 1986) can also be added to diets, but their effects may be proportionately reduced by dilution in the diet.

Biochemical interactions may occur between allelochemicals obtained from plant extracts and components in the artificial diet, masking the effects of the extract or allelochemical (Reese 1983). In addition, most artificial diets are superoptimal, and subtle allellochemical effects may be masked (Rose et al. 1988). For this reason, it is important to assay allelochemicals in diets that closely resemble the nutritional content of the pest insect's host plant, at concentrations in which the allelochemical occurs in fresh plant tissue.

Techniques also exist for the collection of allelochemicals from intact tissues of growing plants. Volatiles emitted by plants can be collected directly from the air by absorption onto porous polymers such as Porapak Q®

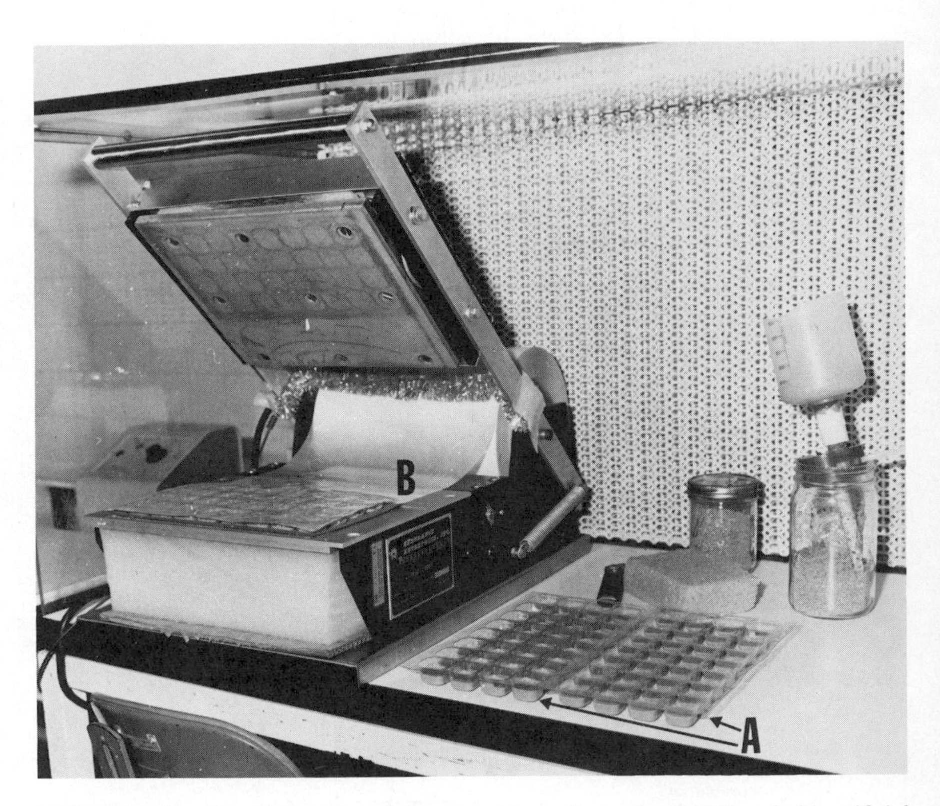

FIGURE 6.15 Plastic insect rearing trays with preformed wells (A) to hold artificial diet and insects. Cover (B) is perforated heavy-duty saran plastic that is heat-sealed to the rearing tray. (Courtesy F. M. Davis and the U.S. Department of Agriculture.)

(ethyl–vinylbenzone–divinyl benzene copolymer) or Tennax (2,6-diphenyl-*p*-phenylene oxide). The adsorbent material can be secured in the stem of a glass funnel with glass wool wadding and the funnel suspended in or over plant foliage. Air is pulled through the funnel with a vacuum pump and the plant volatiles are trapped on the adsorbent (Fig. 6.13). After collection, the adsorbent is removed and the volatile compounds can be eluted with several different nonpolar organic solvents. The extract is then concentrated by evaporation with nitrogen, applied to an inert material, and bioassayed for attraction or repellency in an olfactometer (see Section 6.3.2.3). Methods also exist for collection of allelochemicals from intact plant root systems using various types of adsorbent resins (Tang 1986) (Fig. 6.14). Collection of allelochemicals is achieved by cycling nutrient solutions through the plant root system several times to accumulate allelochemicals produced by plant roots. Extraction of these allelochemicals from the adsorbent resins is similar to that for volatile allelochemicals.

Containers used to evaluate allelochemicals should be small and easy to handle, and should afford the researcher the ability to view the bioassay in progress. Common test containers include the 30 ml polystyrene plastic insect rearing cup, trays with preformed wells for diet and insects (Fig. 6.15), and plastic or glass petri dishes with quadrant divisions (Fig. 6.11). Though they lack humidity control, paper ice cream cartons, with ventilated inserts in the lids, may also be suitable for use. For an in-depth discussion of insect feeding bioassays, see the review of Lewis and van Emden (1986).

6.3 MEASUREMENTS OF RESISTANCE

6.3.1 Plant Measurements

6.3.1.1 Direct Insect Feeding Injury

Measurements of insect damage to plants are usually more useful than measurements of insect growth or population development on plants, because reduced plant damage and the resulting increases in yield or quality are the ultimate goals of most crop improvement programs. Often, measurements of yield reduction indicate direct insect feeding injury in plants. Soft x-ray photography has been used to determine the effect of insect infestation on cotton seed quality (George et al. 1983) and on maize root growth (Villani & Gould 1986). Tissue damage in plants can also be determined by measuring the incidence of tissue necrosis, fruit abscission, and stem damage. The severity of

TABLE 6.3 Rating Scale Used to Evaluate Maize Genotypes for Leaf Feeding Resistance to *Heliothis, Ostrinia, Chilo, Spodoptera,* and *Diatraea*[a]

Resistance Category	Rating	Description
Highly resistant	1	No damage or few pinholes
Resistant	2	Few shot holes on several leaves
	3	Shot holes on several leaves
Intermediately resistant	4	Several leaves with shot holes and a few long lesions
	5	Several leaves with long lesions
	6	Several leaves with lesions < 2.5 cm
	7	Long lesions common on $\frac{1}{2}$ of leaves
Susceptible	8	Long lesions common on $\frac{1}{2}$–$\frac{2}{3}$ of leaves
	9	Most leaves with long lesions

[a]From Guthrie et al. (1960, 1978) and Mihm (1983a, b).

TABLE 6.4 Rating Scales Used to Evaluate Rice for Resistance to Common Insect Pests[a]

Score	Striped Stem Borer (% dead heart)	Gall Midge (% galls)	Leafhopper/ Plant hopper Damage	Whorl Maggot Damage
0	0	0	None	No lesions
1	1–10	< 1	Slight	Pinhead lesions
3	11–25	1–5	Leaves 1 and 2 yellow	Lesions ca. 1 cm long
5	26–40	6–15	Plants stunted; $> \frac{1}{2}$ leaves yellow	Lesions > 1 cm, but $> \frac{1}{2}$ total leaf
7	41–60	16–50	$> \frac{1}{2}$ plants dead; $\frac{1}{2}$ severely stunted and wilted	Lesions on ca. $\frac{1}{2}$ leaf
9	61–100	51–100	All plants dead	Large lesions on $\frac{1}{2}$ leaf, leaf broken

[a]From Heinrichs et al. (1985).

TABLE 6.5 Rating Scales Used to Measure Insect Resistance in Sorghum[a]

Score	Greenbug Damage[b]	Sorghum Midge (% damage)
0	—	0
1	No red spots on leaves	1–10
2	Red spots on leaves	11–20
3	Part of one leaf dead	21–30
4	One leaf dead	31–40
5	Two leaves dead	41–50
6	Four leaves dead	51–60
7	Six leaves dead	61–70
8	Eight leaves dead	71–80
9	Entire plant dead	81–90
10	—	91–100

[a]From Johnson and Teetes (1979).
[b]Estimates of greenbug population size should also be taken if possible.

virus-related stunting, yellowing, or curling can indicate resistance to the virus vector or resistance to the virus itself. Measurements of the cosmetic grade of fruits and vegetables can also be used to measure the effect of insect damage on the aesthetic value of produce.

Insect defoliation to plants is routinely determined by rating scales that make use of visual estimates of plant damage based on percentages or numerical ratings. Several such scales are used to evaluate damage by important insect pests of maize (Table 6.3), rice (Table 6.4), and sorghum (Table 6.5). Leaf area meters are also commonly used to assess differences in insect defoliation of different plants cultivars (Kogan & Goeden 1969).

Indirect feeding injury measurements related to plant growth such as photosynthetic, transpiration, and respiratory rates can also be recorded, although they are somewhat removed from reductions in plant biomass yield.

6.3.1.2 Simulated Feeding Injury

Insect feeding injury can be simulated by mechanical defoliation. However, plants respond somewhat differently to artificial defoliation and to actual insect tissue removal. Therefore, the relationship between the results of artificial and natural defoliation should be determined before accepting results based on artificial defoliation exclusively.

Feeding injury can also be simulated by injection or application of phytotoxic insect secretions into or onto plant tissues. The application of crude extract of greenbug, *Schizaphis graminum* Rondani, to sorghum foliage causes plant reactions similar to those resulting from actual greenbug feeding (Reese & Schmidt unpubl. manuscript). This technique holds great promise as a means of evaluating plant materials.

6.3.1.3 Correlation of Plant Factors with Resistance

Although a thorough knowledge of the actual cause of resistance may not be essential for the development of resistant cultivars, this information may target specific phenotypic characters that can be monitored during the breeding and selection process. If resistance is chemically or morphologically based, concentrations of allelochemicals or the density and size of morphological structures present in tissues of resistant cultivars can be evaluated, allowing a more rapid determination of potentially resistant plant materials.

The evaluation of plant allelochemical and morphological factors also removes the variation due to the test insect until a later stage of study, when results can be confirmed in replicated field experiments under insect infestation. However, the demonstration of allellochemical or morphological differences between resistant and susceptible plants does not always conclusively demonstrate that these factors are involved in conditioning insect resistance.

In a few instances, some of the physical and allelochemical resistance factors described in Chapters 2 and 3 have been used to monitor for insect resistance in different crop plants. Quantitative differences in the trichome-based insect resistance in wheat and alfalfa have been monitored (Hoxie et al. 1975, Kitch et al. 1985). Robinson et al. (1982) developed an accurate, efficient thin layer chromatography technique to identify maize lines with high concentrations of 6-methoxybenzoxazolinone (MBOA) for resistance to the European corn borer. Ultraviolet fluorescence of organic acids has also been adapted to monitor insect resistance in plants. Cole (1987) demonstrated that the fluorescence of lettuce roots resistant to the lettuce root aphid, *Pemphigus bursarius* L., is more intense than that of susceptible roots, owing to higher quantities of isochlorogenic acid (see Section 3.2.1.1). Allellochemical profiles may also be of use by plant breeders to improve cultivars for insect resistance in cotton (Hedin et al. 1983), maize (Hedin et al. 1984), strawberry (Hamilton-Kemp et al. 1988), and tomato (Andersson et al 1980, Quiros et al. 1977).

6.3.2 Insect Measurements

Measurements of insect population development and behavior provide important supplemental information to plant measurements of resistance. Such insect measurements are of primary importance in determining the existence of antibiosis, antixenosis, and/or tolerance.

6.3.2.1 Sampling Insect Populations

Insect populations should be sampled at the plant site where insect damage occurs, during the phenological development of the plant when the pest insect is normally present, and at a time of day when the insect is normally active on the plant. Populations of nonmobile insects may be estimated visually, but this method is subject to error because of the variations in canopy size between different cultivars of crop plants. Shaking or beating plant foliage to dislodge insects onto a ground cloth for counting or collection is a more accurate method of population estimation. More mobile or active insects can be better sampled with a sweep net or a mobile vacuum collection machine (D-Vac), or by anesthetizing caged insects with carbon dioxide. The vacuum collection method is usually less damaging than the sweep net method, but it is most effective in collecting light-bodied insects. Pitfall traps, sticky traps, light traps, and pheromone traps may also be used to measure insect population density, but these measurements are at best only indirect, because they are competing with the test plants for insect attractiveness.

6.3.2.2 Measurements of Insect Feeding and Development

Insect development can be monitored to determine if antibiotic and antixenotic effects are exhibited by insects confined to the foliage of resistant cultivars. Several measures of insect metabolic efficiency can also be determined, using various nutritional indices (Waldbauer 1968). The Consumption Index (CI), is calculated as the weight of food eaten/(mean weight of larvae during testing/test duration). Approximate digestibility (AD), is calculated as (the weight of food eaten – the weight of feces)/the weight of food eaten. The efficiency of conversion of ingested food (ECI) is calculated as the weight gain of larvae/the weight of food eaten. The efficiency of conversion of digested food (ECD) is calculated as the weight gain of larvae/(weight of food eaten – weight of feces). All measurements can be accurately determined using spreadsheet software for the microcomputer (Schmidt & Reese 1986).

The CI indicates whether antixenotic qualities are present in the resistant cultivar, because consumption depends on positive gustatory stimulation in the early stages of feeding. AD, ECI, and ECD are all indicators of potential antibiotic effects of a resistant cultivar, because each measures metabolic processes that affect insect nutritional physiology. Nutritional index measurements have been shown to give the greatest precision when insects consume at least 80% of the available food offered in the experiment (Schmidt & Reese 1986). These measurements have been used to determine resistance mechanisms in cotton (Mulrooney et al. 1985, Montandon et al. 1986, Shaver et al. 1970) and soybean (Reynolds et al. 1984). For a more in-depth discussion of the use of nutritional indices, see Reese (1978).

Nutritional index values are difficult to determine for feeding by Heteroptera and Homoptera. However, Khan and Saxena (1984a) have devel-

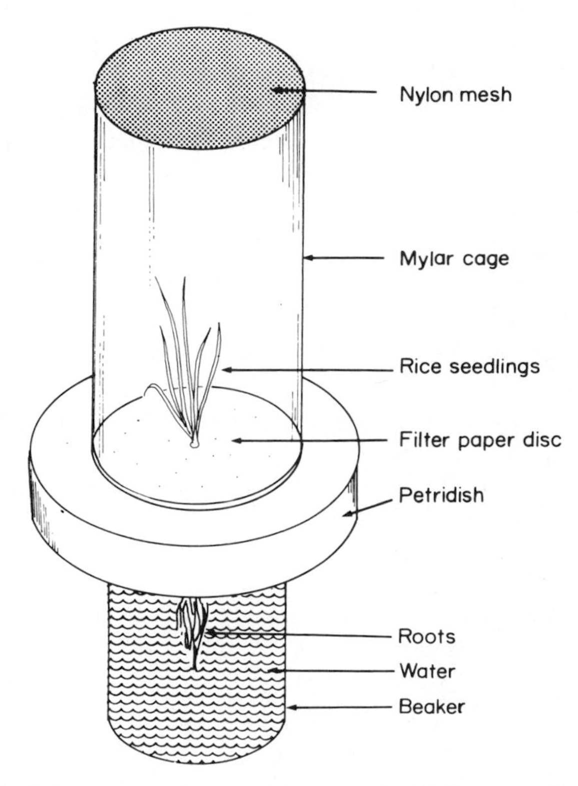

FIGURE 6.16 Apparatus used to monitor green rice leafhopper, *Nephotettix cincticeps* (Uhler), honeydew production on treated filter paper after ingestion of plant sap from resistant and susceptible rice cultivars. (From Khan and Saxena 1984a. Reprinted with permission from J. Econ. Entomol. 77:550–552. Copyright 1984, Entomological Society of America.)

oped a technique to monitor feeding by the green rice leafhopper, *Nephotettix cincticeps* (Uhler), after uptake by rice plants of a dye (Fig. 6.16). The excretion of red honeydew by hoppers feeding on treated resistant seedlings indicates xylem feeding. Conversely, hoppers fed treated susceptible seedlings excrete clear honeydew, indicating phloem feeding. In related research, Pathak and Heinrichs (1982) developed a technique to monitor feeding of the brown plant hopper, *Nilaparvata lugens* Stal, on rice using filter paper treated with bromocresol green. Where drops of honeydew containing amino acids contact the treated filter paper, a colorometric reaction occurs, and dark blue spots appear. The area of the spots is strongly correlated to the weight of actual honeydew produced during feeding on both resistant and susceptible cultivars. The review of Chansigaud and Strebler (1986) offers an extensive review of the methods for studying the effects of allelochemicals on the feeding behavior of Homoptera.

6.3.2.3 Measurements of Insect Behavior

Several techniques used to quantify insect behavior may also be of use in the determination of plant antixenosis. The olfactory responses of insects to volatile allelochemical stimuli have been observed since the beginning of the twentieth century, when Barrows (1907) and McIndoo (1926) developed the first *olfactometers*. Simple two-choice (Y-tube) olfactometers or multichoice devices have been used to measure the response of several species of insects to volatile plant allelochemicals (Ascoli & Albert 1985, Brewer et al. 1983, Byrne & Steinhauer 1966, Panella et al. 1974, Smith et al. 1976, Wilson & Bean 1959) (Fig. 6.17). More complex olfactometers, such as a low-speed wind tunnel (Visser 1976) and an olfactometer room (Payne et al. 1976), have also been used to measure insect olfactory responses to plant volatiles. Such information may enable plant breeders to select cultivars with reduced levels of plant attractants or increased levels of plant repellents. For a complete discussion of insect olfactometers see comments by Finch (1986) and Visser (1986).

Insect behavior can also be measured electrophysiologically, to indicate the effects of the resistant plant cultivar at the insect organ level. *Electroantennograms* (EAGs) are measures of the response of olfactory receptors (commonly located on antennal sensilla) to plant stimuli. Perception of positive (attractant) stimuli elicits hyperpolarizing nerve potentials, whereas perception of negative (repellent) stimuli elicits hypopolarizing nerve potentials (Fig. 6.18). These electrical potentials are transduced by a clamping preamplifier and can be placed in computer storage or displayed on either an oscilloscope or strip chart recorder. Electroantennograms are routinely obtained by splitting the outlet

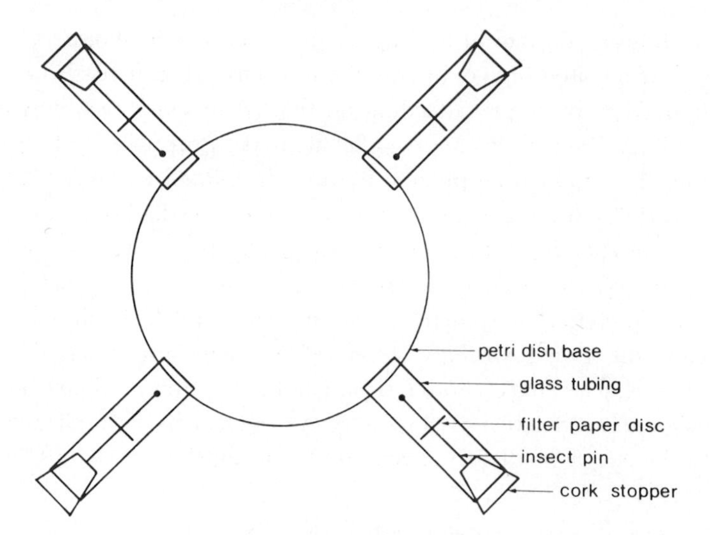

FIGURE 6.17 Simple static flow olfactometer for evaluating insect responses to volatiles applied to filter paper discs.

from a gas chromatograph column between the chromatograph detector and the insect antenna. Tentative assignments of the effects of individual components of plant aroma on insect olfactory perception can be made by comparing the retention times of plant volatiles separated chromatographically with the insect EAG response.

The response of insect contact chemoreceptors to various allelochemicals can also be determined electrophysiologically. The allelochemical stimulus is applied as a vapor or as a solution and the resulting positive (stimulant) or negative (deterrent) nerve potentials are recorded and observed. Frazier and Hanson (1986) provide a detailed explanation and discussion of the use of electrophysiological recording techniques to monitor insect chemosensory responses.

Electroretinograms (ERGs) measure an insect's visual response spectrum and can provide information concerning an insect's perception of both monochromatic light and color (Fig. 6.19). Techniques for determining ERGs of several different insects have been developed by Agee (1977). Electroretinogram data may be useful to plant breeders by providing them data that will enable the development and production of plant cultivars with pigmentation patterns outside of or abnormal to an insect's normal color perception range (see Chapter 2).

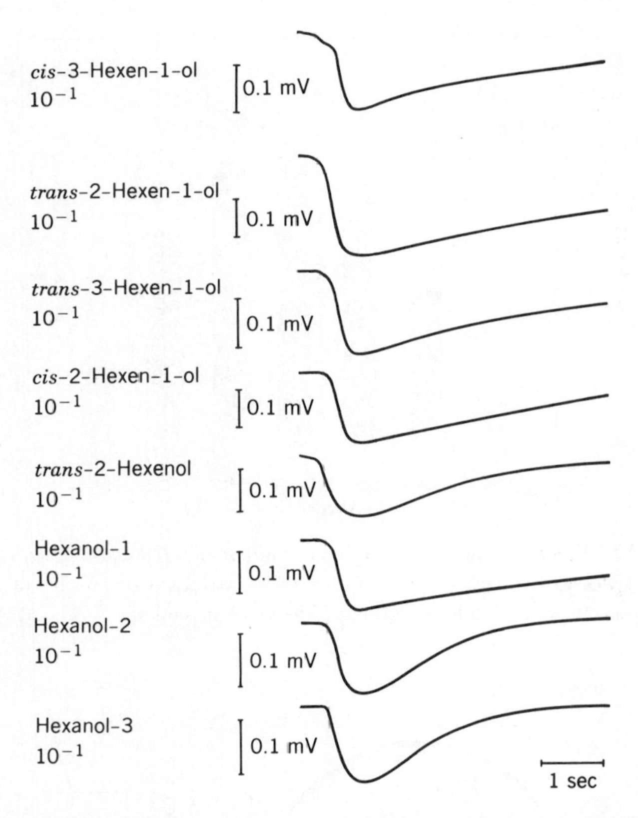

FIGURE 6.18 Electroantennograms of Colorado potato beetles, *Leptinotarsa decemlineata* (Say) responding to green leaf volatiles of potato foliage and their isomers. Vertical marker denotes intensity of receptor response, horizontal marker denotes duration of response. (From Visser 1979. Reprinted with the permission of D. W. Junk Publishing Co.)

An *electronic feeding monitor*, developed by McLean and Kinsey (1966) and McLean and Weigt (1968), has been used by several investigators to quantify the feeding activity of various species of Homoptera. Using this device, a small electrical current is passed across the insect and plant, both of which are wired to a recording device such as a strip chart recorder or an oscilloscope. Feeding activity is detected when the insect feeding stylets penetrate the plant tissue at various depths, causing a change in the electrical conductance by the plant tissues (Fig. 6.20). These changes are converted electronically and displayed as feeding wave forms (Fig. 6.21). Differences in the type of wave form produced

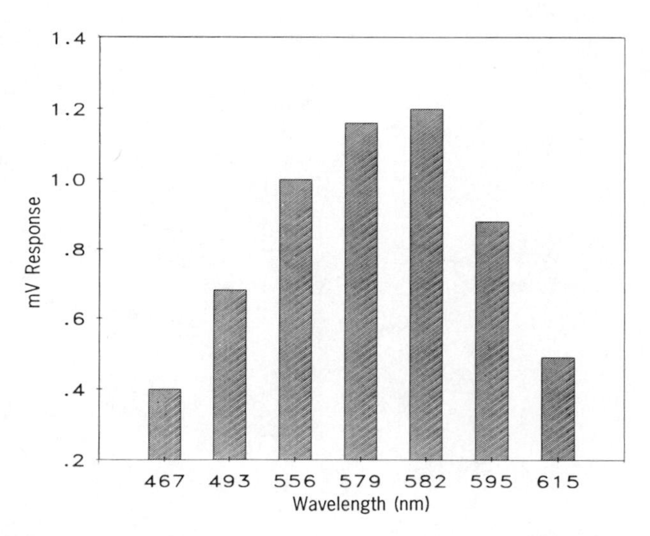

FIGURE 6.19 Electroretinogram of tobacco budworm, *Heliothis virescens* (F.), moth. Summed receptor cell generator potential (mV) at various wavelengths of visible light (500–550 nm, yellow; 550–590 nm green; 590–610 nm, violet).

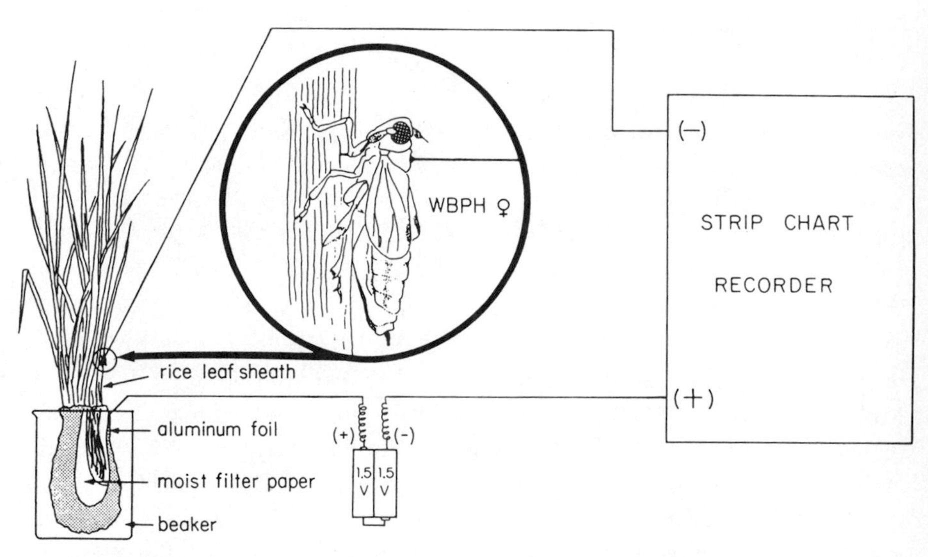

FIGURE 6.20 Schematic diagram of technique for electronically monitoring feeding of the white-backed plant hopper, *Sogatella furcifera* Horvath, on rice plants. (From Khan & Saxena 1984b. Reprinted with permission from J. Econ. Entomol. 77:1479–1482. Copyright 1984, Entomological Society of America.)

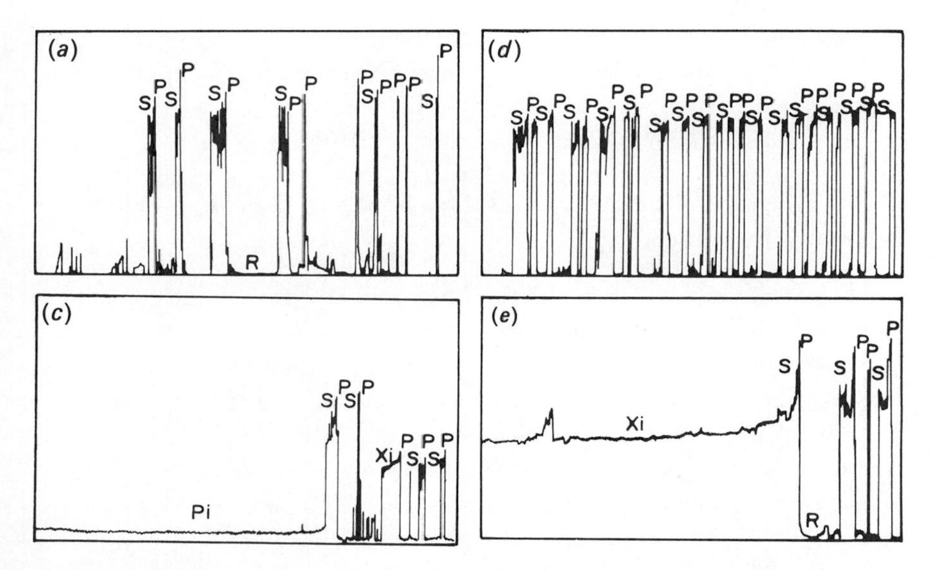

FIGURE 6.21 Wave forms of green rice leafhoppers, *Nephotettix virescens* (Distant), feeding on susceptible (a, c) and resistant (d, e) rice cultivars, as determined by an electronic monitor. (P) Probe, (S) salivation, (Pi) phloem ingestion, (Xi) xylem ingestion, (R) rest. Each record is approximately 4 min long. (From Khan & Saxena 1985. Reprinted with permission from J. Econ. Entomol. 78:583–587. Copyright 1985, Entomological Society of America.)

during insect feeding indicate the frequency of feeding and differences in food source (plant xylem or phloem).

Electronic monitoring has been used in investigations of plant resistance to aphid pests of alfalfa, sorghum strawberry, musk melon, and various *Solanum* species (Holbrook 1980, Campbell et al. 1982, Kennedy et al. 1978, Nielson & Don 1974, Shanks & Chase 1976). Electronic monitoring has also revealed differences in the feeding responses of plant hoppers on resistant and susceptible rice cultivars (Khan & Saxena 1984b), Velusamy & Heinrichs 1986). A complete system for recording, storage, and analysis of feeding data measured by the electronic monitoring technique is described by Kimsey and McLean (1987). Tarn and Adams (1982) have prepared a review of the history, development, and use of the electronic feeding monitor technique.

The observation and analysis of insect behavior patterns over prolonged periods of time has been aided by the development of compact, inexpensive video tape recording and playback equipment. Such technology makes possible the nearly continuous monitoring of different types of behavior on the plant.

With adaptations for image magnification, video recording of insect behavior on resistant plants can be used in conjunction with the electronic feeding monitor.

6.3.3 Use of Biotechnological Methods to Develop Plants Resistant to Insects

Society has been several revolutions of thought or "Ages" as they have been recorded in history. The Renaissance and the Industrial Revolution of the past, along with the more recent Atomic Age and Age of Information Technology, have all shaped the world we live in. Research on plant resistance to insects and much of all of agricultural research are currently involved in the advent of the Age of Molecular Biology. The result of this process is being funneled into advanced biotechnology applications. In many ways, the achievements of the agricultural Green Revolution of the 1960s have now given way to a "Gene Revolution" in agricultural research in the 1980s.

New discoveries in crop plant molecular genetics are now taking place at a rapid rate, making projections about the future of the molecular bases of plant resistance to insects difficult. The following discussions of the uses of cell and tissue culture, genetically engineered plants, and electrophoretic techniques in plant resistance to insect research may well be outdated by the time they are in print. Nevertheless, these brief reviews are intended to define and describe the use of new technologies in the development of plant resistance to insects.

6.3.3.1 Electrophoretic Techniques

Electrophoretic separation techniques are now commonly used by entomologists in many areas of specialization to study differences in insects at the subpopulation level. As indicated in Chapter 9, electrophoresis is being used to differentiate biotypes of the brown plant hopper, *Nilaparvata lugens* Stal, (Sogawa 1978, Saxena 1984) and the greenbug (Kindler et al. unpubl. manuscript). Schukle and Murdock (1983) have also used agar gel radial diffusion, a related technique, to assay plant inhibitors found in the gut tissues of Hessian fly larvae, as possible mechanisms by which fly larvae overcome resistant wheat cultivars.

Electrophoresis is also in use by plant breeders to characterize cultivars of barley, oats, rice, and wheat (Guo et al. 1985, Kiang & Wu 1979, Lookhart 1985, Lookhart et al. 1982, McCausland & Wrigley 1977). Medina Filho and Stevens (1980) used electrophoresis to detect a linkage in tomato genes expressing nematode resistance and acid phosphotase. Thus nematode-resistant cultivars are identified by the presence of an acid phosphotase electromorph and the amount of time required to identify resistance is decreased by approximately 30%. Schukle and Dhawale (unpubl. manuscript) have used electrophoresis to detect differences in the foliar proteins of Hessian fly-infested wheat plants. Electrophoretic techniques are at a very early stage of utilization in the identification of plants with insect resistance.

Obviously, these techniques may not be useful in all plant–insect interactions and they may not be useful as a primary resistance screening technique. One obvious immediate use of electrophoresis is to serve as a method of cataloging the electrophoretic banding patterns of known sources of insect-resistant crop plant material. These patterns can then be used to build a battery of diagnostic tests by which groups of plant material can be evaluated. Various banding patterns that reflect different types and levels of resistance can then be used to indicate potential insect resistance in plant material under evaluation. There is also a future potential to study insect resistance in plants using enzyme immunoassay technology. In medicine, ultrasensitive, specific, and rapid screening techniques exist for the immunological detection of antigens and antibodies (Monroe 1985). Because these are protein-based measurements, the possibility exists for the development of techniques to monitor insect resistance in plant material rapidly by enzyme-linked immunoassays. This technology has recently been introduced into plant disease management programs.

6.3.3.2 Plant Tissue Culture and Cell Culture

Callus (undifferentiated) plant tissue has many potential uses in developing insect-resistant crop plants. Plants derived from callus tissue may possess sufficient genetic variability to contain insect resistance. Such variation in corn, sugarcane, rice, and oat callus has been used to select for disease resistance, agrochemical tolerance, and improved agronomic traits. Attempts to identify insect resistance in somaclonal variants (plants generated from the variation among cultured cells or tissue) are still in an early stage of development, and have not identified differences. White and Irvine (1987) produced 2000 somaclonal variants of sugarcane callus tissue from a cultivar susceptible to the sugarcane borer, *Diatraea saccharalis* (F.). Regenerated plants grown and infested in field plots exhibited random variation in the amount of borer damage incurred; increased levels of borer resistance were detected.

Callus of insect-resistant plant material does appear to exhibit insect resistance. Maize callus tissue exhibits levels of resistance to the fall armyworm,

Spodoptera frugiperda (J. E. Smith), the southwestern corn borer, *Diatraea grandiosella* Dyar, and the corn earworm, *Heliothis zea* Boddie, similar to whole plant foliage (Williams & Davis 1985, Williams et al. 1985, 1987a, b). Callus tissues from fall armyworm-resistant Bermuda grass cultivars also exhibit a resistance reaction similar to whole plant foliage (Croughan & Quisenberry 1989).

Callus tissues may offer additional uses for the plant resistance researcher. If the plant material that is to be evaluated is not adapted to local field growing conditions, callus tissue can be used in place of plants in resistance evaluations. Callus from susceptible plant material can be used as insect rearing medium or as a biological reactor to produce insect rearing ingredients more closely suited to a specific insect's plant diet. Detailed analysis of callus tissues may also provide information on the chemical composition of more nutritionally useful artificial insect diets. Callus tissue also forms the basis of the concept of artificial seeds (Fujii et al. 1987), which consist of somatic embryos in an artificial gel such as calcium alginate. Artificial seeds have been developed for alfalfa and several vegetable crops, offering the possibility of serving as a medium for insect-resistant embryos that can be mass-produced with rigid genetic uniformity.

Plant callus tissues may also be produced in cellular suspension. Such systems can produce allelochemicals and insecticides such as alkaloids, benzoquinones, furarnocoumarins, proteins, and tannins (Fowler 1983). As with the production of insect nutrients, the possibilities exist for entomologists and phytochemists to use plant cell culture to obtain allelochemicals more efficiently than by production of whole plants. Kogan and Schroeder (unpubl. manuscript) are using soybean cell culture to study the allelochemistry of induced insect resistance in soybean foliage.

Cell culture may also be used as a means of directly selecting insect-resistant plants from cell populations. Salt or aluminum tolerant carrot, rice, sugarcane, taro, and tomato plants have all been developed by applying either ion directly to the cell culture (Liu & Yeh 1982, Meredith 1978, Nyman et al. 1983, Ojima & Ohira 1982, Wong & Woo 1983). Cao-Jun & Rush (unpubl. manuscript) have used cell culture selection to develop *Rhizoctonia* fungi resistant rice plants, by applying phenylacetic acid (the fungal toxin) to cell culture. Several Heteroptera inject plant cell-digesting enzymes into plants during feeding (Miles 1972, Teetes 1980). The greenbug, *Schizaphis graminum Rondani*, the yellow sugarcane aphid, *Sipha flava* Forbes, and the Russian wheat aphid, *Diuraphis noxia* (Mordvilko), also inject enzymatic toxins into host tissues. If the toxins from these aphids can be identified, they may be used in a manner similar to the *Rhizoctinia* fungal toxin to treat plant cell cultures and select insect-resistant plant types. Schukle et al. (1985) have partially characterized the enzymes of Hessian fly, *Mayetiola destructor* (Say), larvae in order to use these enzymes in cell culture selection in wheat for fly resistance.

6.3.3.3 Resistance to Insects in Genetically Transformed Plants

Recent research results demonstrate that insect resistance can be obtained by nonconventional plant breeding techniques. Fischoff et al. (1987) developed transgenic tomato plants with resistance to several foliage feeding Lepidoptera derived from the HD-1-delta-endotoxin (tox) gene. This gene encodes for the production of a crystal protein toxin from *Bacillus thuringiensis* var. *kurstaki*. After determination of the gene structure by cloning and hybridization, truncated genes were incorporated into tomato callus via the Ti plasmid of the bacterial vector *Agrobacterium tumefaciens*. Foliage from plants surviving the process of regeneration from callus tissue was fed to larvae of the tobacco hornworm, *Manduca sexta* (L.), tobacco budworm, *Heliothis virescens* (F.), and the corn earworm in laboratory bioassays. Larval mortality ranged from 50 to 100%. Comparable results have been observed in transgenic tobacco plants produced under similar conditions that are resistant to the tobacco hornworm (Adang et al. 1986, Vaeck et al. 1987).

Though gene transfer has not progressed as far in monocot plants, partial transfer of the tox gene into maize plants has been accomplished by inserting the gene into *Pseudomonas* maize root bacteria. Stock et al. (1986) first transferred the tox gene into *Pseudomonas cepacia*, which was used to colonize tobacco leaves. The colonized foliage is toxic when fed to tobacco hornworm larvae. Obukowicz et al. (1986, 1987) successfully inserted the tox gene into the chromosomes of two *Pseudomonas fluoresans* strains that directly colonize maize roots. Bioassay of colonized plant material awaits approval by the U.S. Environmental Protection Agency. However, tobacco hornworm harvae suffer 100% mortality after consumption of artificial diet amended with bacterial isolates containing the tox gene.

Other methods of gene transfer in monocots such as electroporation and protoplast transformation have recently become a reality (Fromm et al. 1986, Lorz et al. 1985, Werr & Lorz 1986). An RNA virus that acts as a gene expression vector in monocot plant cells has been identified (French et al. 1986). The use of one or more of the above techniques will most likely result in the development of transgenic insect-resistant monocots in the near future.

To date, insecticidal activity of *B. thuringiensis* (*B.t.*) strains such as *B.t. kurstaki* HD-1 has been limited to Lepidoptera and Diptera. However, new *B.t.*

strains with Coleoptera-specific mortality have recenty been identified that have great promise for use in the development of transgenic crop plants with pest Coleoptera resistance. Herrnstadt et al. (1986) isolated the *B.t.* strain *Bacillus thuringiensis san diego*, that is toxic to the boll weevil, *Anthomomus grandis* (Boheman), the Colorado potato beetle, *Leptinotarsa decemlineata* (Say), and the elm leaf beetle, *Pyrrhalta luteola* (Muller). The discovery of this new *B.t.* strain provides a great opportunity to identify the genes responsible for its toxicity and eventually to encode these genes into transgenic crop plants.

Genetically altered cultivars of several types of crops will likely be developed and released in the near future. For this reason, there is a real need for development of cultivar release strategies that avoid the development of resistance-breaking insect biotypes. Such strategies are necessary because of the high potential that exists for the selection of *B.t.* resistant pest populations when crops with *B.t.* toxic genes are released for production. Such strategies are especially necessary for highly polyphagous species of pest insects such as migratory Lepidoptera and Coleoptera that will be exposed to the toxin in several different crops in the same agroecosystem.

Restriction enzyme techniques developed by molecular biologists are also being used to determine the gene loci responsible for plant resistance to insects. Restriction fragment length polymorphisms (RFLP), differences in the fragment lengths of restriction endonuclease-digested DNA of plant genotypes, can act as genetic markers such as isozyme markers. Differences in fragment lengths, detected by agarose gel electrophoresis, function as alleles of a particular gene locus. The colorimetric absorbance of 2-tridecanone, the allelochemical responsible for insect resistance in *Lycopersicon hirsutum* f. *glabratum*, and the RFLP loci on three different linkage groups among a group of *Lycopersicon* plants varying widely in 2-tridecanone content are correlated (Nienhuis et al. 1987). Thus direct selection for polymorphic marker loci can result in an increased frequency of alleles related to the expression of 2-tridecanone-mediated resistance in tomato. Restriction endonuclease techniques are also in use to map the messenger-RNA genome of resistance genes in wheat cultivars resistant to the Hessian fly (Shukle, unpubl. manuscript).

6.4 CONCLUSIONS

The techniques and devices used to record insect activity on plants or the resulting plant damage are as varied as the particular combination of pest insect and host plant. Techniques such as mass seedling evaluation and mechanical inoculation of plants with artificially reared insects have led to great increases in the speed and efficiency of plant resistance evaluations. As a result, the development of insect-resistant cultivars of several crop plants has occurred at a much faster than normal rate. The adoption of nutritional indices techniques and of accurate, accepted damage rating scales has also increased the degree of accuracy involved in the definition of the levels of insect resistance in several species of crop plants.

Although much progress has been made in developing accurate and efficient techniques to assess plant resistance to insects, more new and improved techniques are still needed. A common problem is the lack of a constant supply of insects with which to conduct research. This problem usually stems from the lack of an artificial diet with which to produce insects or from deficient rearing and handling methods for insects produced on an existing diet. In research with caged insect populations, there is a need to develop cage materials that allow better light transmission and air flow while maintaining sufficiently small pore sizes to contain insects. Improved insect infestation techniques and devices that consume less time and minimize damage to insects from handling, such as the larval Lepidoptera inoculator, will also be very useful. Finally, the development and refinement of standardized rating scales to determine insect damage or defoliation in crops other than those mentioned previously (maize, sorghum, and rice) would greatly facilitate the development of insect cultivars in several additional crop plant species.

In laboratory research, there is also a need for improved knowledge of plant nutrient composition, in order to design artificial insect diets that more accurately reflect the nutrient composition of an insect's host plant. This understanding is critical to the success of determining the true contributions of plant allelochemicals to insect resistance. Most of the artificial diets currently used for laboratory culture of Lepidoptera are superoptimal in nutrient concentration and can easily mask the subtle effects of some allelochemicals. There also continues to be a need for improved analytical techniques, especially microanalytical techniques, to determine allelochemical resistance factors in plants.

As previously indicated, most of the progress in plant resistance to insects has been due to conventional insect evaluation and plant breeding techniques. In the future, accurate and economical techniques of plant and insect evaluation developed from biotechnology will also prove useful, as molecular plant technologies are adapted for use in crop protection. However, problems inherent in the development of insect resistance (nontarget insect susceptibility, potential for development of insect biotypes, and poor agronomic qualities in resistant parent material) will most likely follow the development of resistant cultivars, whether by conventional or transgenic means. New techniques to identify plant resistance to insects, using all available technologies, will be a main element of future crop insect pest management systems.

REFERENCES

Adams, C. M. and E. A. Bernays. 1978. The effect of combinations of deterrents on the feeding behavior of *Locusta migratoria*. Entomol. Exp. Appl. 23:101–109.

Adang, M. J., D. DeBoer, E. Firoozabady, J. D. Kemp, E. Murray, T. A. Rocheleau, K. Rashka, G. Staffeld, C. Stock, D. Sutton, and D. J. Merlo. 1986. Applications of a *Bacillus thuringiensis* crystal protein for insect control. J. Cell Biochem. Suppl. 1OC. Proc. UCLA Symp. Molecular and Cell Biology. Los Angeles, CA, p. 11. Poster J20.

Agee, H. R. 1977. Instrumentation and techniques for measuring the quality of insect vision with the electroretinogram. USDA-ARS-S-162. 13 pp.

Agarwal, R. A., J. P. Singh, and C. B. Tiwari. 1971. Technique for screening of sugarcane varieties resistant to top borer, *Scirpophaga nivella* F. Entomophaga 16:209–220.

Andersson, B. A., R. T. Holman, L. Lundgren, and G. Stenhagen. 1980. Capillary gas chromatograms of leaf volatiles. A possible aid to breeders for pest and disease resistance. J. Agric. Food Chem. 28:985–989.

Ascher, K. R. S. and N. E. Nemny. 1978. Use of foam polyurethane as a carrier for phagostimulant assays with *Spodoptera littoralis* larvae. Entomol. Exp. Appl. 24:546–548.

Ascoli, A. and P. J. Albert. 1985. Orientation behavior of second-instar larvae of eastern budworm, *Choristoneura fumiferana* (Clem.) (Lepidoptera: Tortricidae) in a Y-type olfactometer. J. Chem. Ecol. 11:837–845.

Barrows, W. M. 1907. The reactions of the pomace fly, *Drosophila ampelophila* Loew., to odorous substances. J. Exp. Zool. 4:515–537.

Benedict, J. H. 1983. Methods of evaluating cotton for resistance to the boll weevil. In *Host Plant Resistance Research Methods for Insects, Diseases, Nematodes and Spider Mites in Cotton.* Southern Cooperative Series Bulletin 280. Mississippi Agricultural and Forestry Experiment Station. pp. 19–26.

Berenbaum, M. 1986. Postingestive effects of phytochemicals on insects: On paracelsus and plant products. In J. R. Miller and T. A. Miller (eds.). *Insect–Plant Interactions.* Springer, New York, pp. 121–153.

Brewer, G. J., E. L. Sorensen, and E. K. Horber. 1983. Attractiveness of glandular and simple-haired *Medicago* clones with different degrees of resistance to the alfalfa seed chalcid (Hymenoptera: Eurytomidae) tested in an olfactometer. Environ. Entomol. 12:1504–1508.

Bristow, P. R., R. P. Doss, and R. L. Campbell. 1979. A membrane filter bioassay for studying phagostimulatory materials in leaf extracts. Ann. Entomol. Soc. Am. 72:16–18.

Byers, R. A. and W. A. Kendall. 1982. Effects of plant genotypes and root nodulation on growth and survival of *Sitona* spp. larvae. Environ. Entomol. 11:440–443.

Byrne, H. D. and A. L. Steinhauer. 1966. The attraction of the alfalfa weevil, *Hypera postica* (Coleoptera: Curculionidae), to alfalfa. Ann. Entomol. Soc. Am. 59:303–309.

Caballero, P. 1985. Allelochemicals from soybean affecting *Pseudoplusia includens* (Walker), biology and pheromone from *Chilo plejadellus* Zincken mediating *C. plejadellus* sexual behavior. Ph.D. dissertation. Louisiana State University. Baton Rouge, 81 pp.

Campbell, B. C., D. L. McLean, M. G. Kinsey, K. C. Jones, and D. L. Dreyer. 1982. Probing behavior of the greenbug (*Schizaphis graminum*, biotype C) on resistant and susceptible varieties of sorghum. Entomol. Exp. Appl. 31:140–146.

Chan, B. G., A. C. Waiss, Jr., W. L. Stanley, and A. E. Goodban. 1978. A rapid diet preparation method for antibiotic phytochemical bioassay. J. Econ. Entomol. 71:366–368.

Chansigaud, J. and G. Strebler. 1986. Methods for investigating the mode of action of natural or synthetic substances on the feeding behavior of Homoptera. Agronomie 6:845–856.

Chelliah, S. and E. A. Heinrichs. 1980. Factors affecting insecticide-induced resurgence of the brown planthopper, *Nilaparvata lugens* on rice. Environ. Entomol. 9:773–777.

Cole, R. A. 1987. Intensity of radicle fluorescence as related to the resistance of seedlings of lettuce to the lettuce root aphid and carrot to the carrot fly. Ann. Appl. Biol. 111:629–639.

Croughan, S. S. and S. S. Quisenberry. 1989. Evaluation of cell culture as a screening technique for determining fall armyworm (Lepidopetera: Noctuidae) resistance in bermudagrass. J. Econ. Entomol. 82:232–235.

Davis, F. M. 1980a. A larval dispenser-capper machine for mass rearing the southwestern corn borer. J. Ecol. Entomol. 73:692–693.

Davis, F. M. 1980b. Fall armyworm resistance programs. Fla. Entomol. 63:420–433.

Davis, F. M. 1982. Mechanically removing southwestern corn borer pupae from plastic rearing cups. J. Econ. Entomol. 75:393–395.

Davis, F. M. 1985. Entomological techniques and methodologies used in research programmes on plant resistance to insects. Insect Sci. Appl. 6:391–400.

Davis, F. M. and T. G. Oswalt. 1979. Hand inoculator for dispensing lepidopterous larvae. USDA/ARS AAT-S-9/October, 5 pp.

Davis, F. M. and W. P. Williams. 1980. Southwestern corn borer: Comparison of techniques for infesting corn for plant resistance studies. J. Econ. Entomol. 73:704–706.

Davis, F. M., T. G. Oswalt, and S. S. Ng. 1985. Improved oviposition and egg collection system for the fall armyworm (Lepidoptera: Noctuidae). J. Econ. Entomol. 78:725–729.

Dilday, R. H. 1983. Methods of screening cotton for resistance to *Heliothis* spp. In *Host Plant Resistance Research Methods for Insects, Diseases, Nematodes and Spider Mites in Cotton.* Southern Cooperative Series Bulletin 280. Mississippi Agricultural and Forestry Experiment Station, pp. 26–36.

Dimock, M. B., S. L. LaPointe, and W. M. Tingey. 1986. *Solanum neocardensaii*: A new source of potato resistance to the Colorado potato beetle (Coleoptera: Chrysomelidae). J. Econ. Entomol. 79:1269–1275.

Doss, R. P. and C. H. Shanks, Jr. 1986. Use of membrane filters as a substrate in insect feeding bioassays. Bull. Entomol. Soc. Am. 32:248–249.

Eenink, A. H., F. F. Dieleman, R. Groenwold, P. Aarts, and B. Clerkx. 1984. An instant bioassay for resistance of lettuce to the leaf aphid, *Myzus persicae*. Euphytica 33:825–831.

Finch, S. 1986. Assessing host-plant finding in insects. In J. R. Miller and T. A. Miller (eds.). *Insect–Plant Interactions*. Springer, New York, pp. 23–64.

Fischoff, D. A., K. S. Bowdish, F. J. Perlak, P. G. Marrone, S. M. McMormick, J. G. Niedermeyer, D. A. Duff, K. Kusano- Kretzmer, E. J. Mayer, D. E. Rochester, S. G. Rogers, and R. T. Fraley. 1987. Insect tolerant transgenic tomato plants. Biotechnology 5:807–813.

Fowler, M. W. 1983. Commercial applications and economic aspects of mass plant cell culture. In S. H. Mantell and H. Smith (eds.). *Plant Biotechnology*. Cambridge University Press, New York, pp. 3–37.

Frazier, J. L. and F. E. Hanson. 1986. Electrophysiological recording and analysis of insect chemosensory responses. In J. R. Miller and T. A. Miller (eds.). *Insect–Plant Interactions*. Springer, New York, pp. 285–330.

French, R., M. Janda, and P. Ahlquist. 1986. Bacterial genes inserted in an engineered RNA virus: Efficient expression in monocotyledonous plant cells. Science. 231:1294–1297.

Fromm, M. E., L. P. Taylor, and V. Walbot, 1986. Stable transformation of maize after gene transfer by electroporation. Nature 319:791–793.

Fujii, J. A., D. T. Slade, K. Redenbaugh, and K. A. Walker. 1987. Artificial seeds for plant propagation. T. I. Biotech. 5:335–339.

George, B. W., F. D. Wilson, and R. L. Wilson. 1983. Methods of evaluating cotton for resistance to pink bollworm, cotton leaf perforator, and lygus bugs. In *Host Plant Resistance Research Methods for Insects, Diseases, Nematodes and Spider Mites in Cotton.* Southern Cooperative Series Bulletin 280. Mississippi Agricultural Forestry Experiment Station, pp. 41–45.

Gibson, R. W. 1976. Glandular hairs on *Solanum polyadenium* lessen damage by the Colorado potato beetle. Ann. Appl. Biol. 82:147–150.

Guo, Y.-J., R. Bishop, H. Ferhnstrom, G.-Z. Yu, Y.-N. Lian, & S.-D. Huang. 1985. Classification of Chinese rice varieties by electrofocusing. Cereal Chem. 63:1–3.

Guthrie, W. D., F. F. Dicke, and C. R. Neiswander. 1960. Leaf and sheath feeding resistance to the European corn borer in eight inbred lines of dent corn. Ohio Agric. Exp. Sta. Res. Bull. 860. 38 pp.

Guthrie, W. D., W. A. Russell, G. L. Reed, A. R. Halbauer, and D. F. Cox. 1978. Methods of evaluating maize for sheath-collar-feeding resistance to the European corn borer. Maydica 23:45–54.

Hamilton-Kemp, R. A. Anderson, J. G. Rodriguez, J. H. Longhrin, and C. G. Patterson. 1988. Strawberry foliage headspace vapor components at periods of susceptibility and resistance to *Tetranychus urticae* Koch. J. Chem. Ecol. 14:789–791.

Hedin, P. A., J. N. Jenkins, D. H. Collum, W. H. White, W. L. Parrot, and M. W. MacGowan. 1983. Cyanidin-3-α-glucoside, a newly recognized basis for resistance in cotton to the tobacco budworm, *Heliothis virescens* (Fab.) (Lepidoptera: Noctuidae). Experientia 39:799–801.

Hedin, P. A., F. M. Davis, W. P. Williams, and M. L. Salin. 1984. Possible factors of leaf-feeding resistance in corn to the southeastern corn borer. J. Agric. Food Chem. 32:262–267.

Heinrichs, E. A., F. G. Medrano, and H. R. Rapusas. 1985. *Genetic Evaluation for Insect Resistance in Rice.* International Rice Research Institute, Los Banos, Laguna, Philippines. 356 pp.

Herrnstadt, C., G. C. Soares, E. R. Wilcox, and D. L. Edwards. 1986. A new strain of *Bacillus thuringensis* with activity against Coleopteran insects. Bio/Technol. 4:305–308.

Hill, R. R., Jr. and R. C. Newton. 1972. A method for mass screening alfalfa for meadow spittle bug resistance in the greenhouse during the winter. J. Econ. Entomol. 65:621–623.

Holbrook, F. R. 1980. An index of acceptability of green peach aphids for *Solanum* plant material and for a suspected non-host plant. Am. Potato J. 57:1–6.

Hollay, M. E., C. M. Smith, and J. F. Robinson. 1987. Structure and formation of feeding sheaths of the rice stink bug (*Hemiptera: Pentatomidae*) on rice grains and their association with fungi. Ann. Entomol. Soc. Am. 80:212–216.

Horber, E. 1980. Types and classification of resistance. In F. G. Maxwell and P. R. Jennings (eds.). *Breeding Plants Resistant to Insects.* Wiley, New York, pp. 15–21.

Hoxie, R. P., S. G. Wellso, and J. A. Webster. 1975. Cereal leaf beetle response to wheat trichome length and density. Environ. Entomol. 4:365–370.

Jenkins, J. N., W. L. Parrot, and J. C. McCarty. 1983. Breeding cotton for resistance to the tobacco budworm: Techniques to achieve uniform field infestations. In *Host Plant Resistance Research Methods for Insects, Diseases, Nematodes and Spider Mites in Cotton.* Southern Cooperative Series Bulletin 280. Mississippi Agricultural and Forestry Experiment Station, pp. 36–41.

Johnson, J. W. and G. L. Teetes. 1979. Breeding for arthropod resistance in sorghum. In M. K. Harris (eds.). *Biology and Breeding for Resistance to Arthropods and Pathogens on Agricultural Plants.* TAMU Publ. MP-1451. pp. 168–180.

Kennedy, G. G., D. L. McLean, and M. G. Kinsey. 1978. Probing behaviour of *Aphis gossyppi* on resistant and susceptible muskmelon. J. Econ. Entomol. 71: 13–16.

Khan, Z. R. and R. C. Saxena. 1984a. Techniques for demonstrating phloem or xylem feeding by leafhoppers (Homoptera: Cicadellidae) and planthoppers (Homoptera: Delphacidae) in rice plant. J. Econ. Entomol. 77: 550–552.

Khan, Z. R. and R. C. Saxena. 1984b. Electronically recorded waveforms associated with the feeding behavior of *Sogatella furcifera* (Homoptera: Delphacidae) on susceptible and resistant rice varieties. J. Econ. Entomol. 77: 1479–1482.

Khan, Z. R. and R. C. Saxena. 1985. Mode of feeding and growth of *Nephotettix virescens* (Homoptera: Cicadellidae) on selected resistant and susceptible rice varieties. J. Econ. Entomol. 78: 583–587.

Khan, Z. R. and R. C. Saxena. 1986. Effect of steam distillate extracts of resistant and susceptible rice cultivars on behavior of *Sorgatella furcifera* (Homoptera: Delphacidae). J. Econ. Entomol. 79: 928–935.

Khan, Z. R., J. T. Ward, and D. M. Norris. 1986. Role of trichomes in soybean resistance to cabbage looper, *Trichoplusia ni.* Entomol. Exp. Appl. 42: 109–117.

Kiang, Y.-T. and L. Wu. 1979. Genetic studies of esterases on the Taiwan wild rice population and cultivated rice. Bot. Bull. Acad. Sinica 20: 103–106.

Kimsey, R. B. and D. L. McLean. 1987. Versatile electronic measurement system for studying probing and feeding behavior of piercing and sucking insects. Ann. Entomol. Soc. Am. 80: 118–192.

Kishaba, A. N. and G. R. Manglitz. 1965. Non-preference as a mechanism of sweetclover and alfalfa resistance to the sweetclover aphid and spotted alfalfa aphid. J. Econ. Entomol. 58: 566–569.

Kitch, L. W., R. W. Shade, W. E. Nyquist, and J. D. Axtell. 1985. Inheritance of density of erect glandular trichomes in the genus *Medicago.* Crop Sci. 25: 607–611.

Kogan, M. and R. D. Goeden. 1969. A photometric technique for quantitive evaluation of feeding preferences of phytophagous insects. Ann. Entomol. Am. 62: 319–322.

Lambert, L. 1984. An improved screen-cage design for use in plant and insect research. Agron. J. 76: 168–170.

Leigh, T. F. 1983. Research methods for cotton resistance to spider mites. In *Host Plant Resistance Research Methods for Insects, Diseases, Nematodes and Spider Mites in Cotton.*

Southern Cooperative Series Bulletin 280. Mississippi Agricultural and Forestry Experiment Station, pp. 56–58.

Lewis, A. C. and H. F. van Emden. 1986. Assays for Insect Feeding. In J. R. Miller and T. A. Miller (ed.). *Insect–Plant Interactions.* Springer, New York, pp. 95–119.

Liu, M. and H. Yeh. 1982. Selection of a NaCl tolerant line through stepwise salinized sugarcane cell cultures. In A. Fujiwara (ed.). *Plant Tissue Culture.* Japan Assn. Plant Tissue Culture, Tokyo, pp. 477–478.

Lookhart, G. L. 1985. Identification of oat cultivars by combining polyacrylamide gel electrophoresis and reversed-phase high-performance liquid chromatography. Cereal Chem. 62: 343–350.

Lookhart, G. L., B. L. Jones, S. B. Hall, and K. F. Finney. 1982. An improved method for standardizing polyacrylamide gel electrophoresis of wheat gliadin protein. Cereal Chem. 59: 178.

Lorz, H., B. Baker, and J. Schell. 1985. Gene transfer to cereal cells mediated by protoplast transformation. Mol. Gen. Genet. 199: 178–182.

Martin, F. A., C. A. Richard, and S. D. Hensley. 1975. Host resistance to *Diatraea saccharalis* (F.): Relationship of sugarcane internode hardness to larval damage. Environ. Entomol. 4: 687–688.

McCausland, J. and C. W. Wrigley. 1977. Identification of Australian barley cultivars by laboratory methods; gel electrophoresis and gel isoelectric focusing of the endosperm proteins. Aust. J. Exp. Agric. Anim. Husb. 17: 1020.

McIndoo, N. E. 1926. An insect olfactometer. J. Econ. Entomol. 12: 545–571.

McLean, D. L. and M. G. Kinsey. 1964. A technique for electronically recording aphid feeding and salivation. Nature 202: 1358–1359.

McLean, D. L. and W. A. Weigt. 1968. An electronic measurement system to record aphid salivation and ingestion. Ann. Entomol. Soc. Am. 61: 180–185.

Medina Filho, H. and M. A. Stevens. 1980. Tomato breeding for nematode resistance: Survey of resistant varieties for horticultural characteristics and genotype of acid. Acta Hortic. 100: 383–393.

Meredith. C. P. 1978. Selection and characterization of aluminum-resistant variants from tomato cell cultures. Plant Sci. Lett. 12: 25–34.

Merkle, O. G. and K. J. Starks. 1985. Resistance of wheat to the yellow sugarcane aphid (Homoptera: Aphididae). J. Econ. Entomol. 78: 127–128.

Mihm, J. A. 1983a. Techniques for efficient mass rearing and infestation in screening for host plant resistance to corn earworm, *Heliothis zea.* International Maize and Wheat Improvement Center, El Batan, Mexico. 16 pp.

ibid. 1983b. Efficient mass-rearing and infestation techniques to screen for host plant resistance to fall armyworm, *Spodoptera frugiperda.* International Maize and Wheat Improvement Center, El Batan, Mexico. 16 pp.

ibid. 1983c. Efficient mass rearing and infestation techniques to screen for host plant resistance to maize stem borers, *Diatraea* sp. International Maize and Wheat Improvement Center, El Batan, Mexico. 23 pp.

Mihm, J. A., F. B. Peairs, and A. Ortega. 1978. New procedures for efficient mass production and artificial infestation with lepidopterous pests of maize. CIMMYT Review. International Maize and Wheat Improvement Center, El Batan. Mexico, 138 pp.

Miles, P. W. 1972. The saliva of Hemiptera. Adv. Insect Physiol. 9: 183–255.

Monroe, D. 1985. The solid-phase enzyme-linked immunospot assay: Current and potential applications. BioTechniques May/June: 222–229.

Montandon, R., R. D. Stipanovic, H. J. Williams, W. L. Sterling, and S. B. Vinson. 1986. Nutritional indices and excretion of gosypol by *Alabama argillacae* (Hubner) and *Heliothis virescens* (F.) (Lepidoptera: Noctuidae) fed glanded and glandless cotyledonary cotton leaves. J. Econ. Entomol. 80: 32–36.

Mulrooney, J. E., W. L. Parrott, and J. N. Jenkins. 1985. Nutritional indices of second-instar tobacco budworm larvae (Lepidoptera: Noctuidae) fed different cotton strains. J. Econ. Entomol. 78: 757–761.

Nielson, M. W. 1974. Evaluating spotted alfalfa aphid resistance. In K. K. Barnes (ed.). *Standard Tests to Characterize Pest Resistance in Alfalfa Varieties*. U.S. Dept. Agric. NC-19: 19–20.

Nielson, M. W. and H. Don. 1974. Probing behavior of biotypes of the spotted alfalfa aphid on resistant and susceptible alfalfa clones. Entomol. Exp. Appl. 17: 477–486.

Nienhuis, J., T. Helentjaris, M. Slocum, B. Ruggero, and A. Schaefer. 1987. Restriction fragment length polymorphism analysis of loci associated with insect resistance in tomato. Crop Sci. 27:797–803.

Nyman, L. P., C. J. Gonzales, and J. Arditti. 1983. *In vitro* selection for salt tolerance of taro [*Colocasia esculenta* (L.) Scott. var. *antiquarum*]. Ann. Bot. 51: 229–236.

Obukowicz, M. G., F. J. Perlak, K. Kusano-Kretzmer, E. J. Mayer, S. L. Bolten, and L. S. Watrud. 1986. Tn-5 mediated integration of the delta-endotoxin gene from *Bacillus thuringensis* into the chromosome of root-colonizing pseudomonads. J. Bacteriol. 168: 982–989.

Obukowicz, M. G., F. J. Perlak, S. L. Bolten, K. Kusano-Kretzmer, E. J. Mayer, and L. S. Watrud. 1987. IS5OL as a non-self transposable vector used to integrate the *Bacillus thuringiensis* delta-endotoxin gene into the chromosome of root-colonizing pseudomonads. Gene 51: 91–96.

Ojima, K. and K. Ohira, 1982. Characterization and regeneration of an aluminum-resistant variant from carrot cell cultures. In A. Fujiwara (ed.). *Plant Tissue Culture*. Japan. Assn. Plant Tissue Culture, Tokyo, pp. 475–476.

Ortman, E. E. and T. F. Branson. 1976. Growth pouches for studies of host plant resistance to larvae of corn rootworms. J. Econ. Entomol. 69: 380–382.

Panella, J. S., J. A. Webster, and M. J. Zabik. 1974. Cereal leaf beetle host selection and plant resistance: olfactometer and feeding attractant tests (Coleoptera: Chrysomelidae). J. Kans. Entomol. Soc. 47: 348–357.

Pantoja, A., C. M. Smith, and J. F. Robinson. 1986. Evaluation of rice plant material for resistance to the fall armyworm (Lepidoptera: Noctuidae). J. Econ. Entomol. 79: 1319–1323.

Pathak, P. K. and E. A. Heinrichs. 1982. Bromocresol green indicator for measuring feeding activity of *Nilaparvata lugens* on rice varieties. Philipp. Ent. 5: 195–198.

Pathak, P. K., R. C. Saxena, and E. A. Heinrichs. 1982. Parafilm sachet for measuring honeydew excretion by *Nilaparvata lugens* on rice. J. Econ. Entomol. 75: 194–195.

Payne, T. L., E. R. Hart, L. J. Edson, F. A. McCarty, P. M. Billings, and J. E. Coster. 1976. Olfactometer for assay of behavioral chemicals for the southern pine beetle. *Dendroctonus frontalis* (Coleoptera: Scolytidae). J. Chem. Ecol. 2: 411–419.

Powell, G. S., W. V. Campbell, W. A. Cope, and D. S. Chamblee. 1983. Ladino clover resistance to the clover root curculio (Coleoptera: Curculionidae). J. Econ. Entomol. 76: 264–268.

Quiros, C. F., M. A. Stevens, C. M. Rick, and M. L. Kok-Yokomi. 1977. Resistance in tomato to the pink form of the potato aphid (*Macrosiphum euphorbiae* Thomas): The role of anatomy, epidermal hairs, and foliage composition. J. Am. Soc. Hortic. Sci. 102: 166–177.

Quisenberry, S. S., P. Caballero, and C. M. Smith. 1988. Influence of bermudagrass leaf extracts on development of fall armyworm (Lepidoptera: Noctuidae) larvae. J. Econ. Entomol. 81: 910–913.

Reese, J. C. 1978. Chronic effects of plant allelochemicals on insect nutritional physiology. Entomol. Exp. Appl. 24: 625–631.

Reese, J. C. 1983. Nutrient-allelochemical interactions in host plant resistance. In P. A. Hedin (ed.). *Plant Resistance to Insects*. ACS Symposium Series No. 208. American Chemical Society, Washington DC, pp. 231–243.

Renwick, J. A. A. 1983. Nonpreference mechanisms: Plant characteristics influencing insect behavior. In P. A. Hedin (ed.). *Plant Resistance to Insects*. ACS Symposium Series No. 208. American Chemical Society, Washington DC, pp. 199–213.

Reynolds, G. W., C. M. Smith, and K. M. Kester. 1984. Reductions in consumption, utilization, and growth rate of soybean looper (Lepidoptera: Noctuidae) larvae fed foliage of soybean genotype PI 227687. J. Econ. Entomol. 77: 1371–1375.

Robinson, J. F., A. Klun, W. D. Guthrie, and T. A. Brindley. 1982. European corn borer leaf feeding resistance: A simplified technique for determining relative differences in concentrations of 6-methoxy-benzoxazolinone (Lepidoptera: Pyralidae). J. Kans. Entomol. Soc. 55: 297–301.

Rose, R. L., T. C. Sparks, and C. M. Smith. 1988. Insecticide toxicity to the soybean looper and the velvetbean caterpiller (Lepidoptera: Noctuidae) as influenced by feeding on resistant soybean (PI227687) leaves and coumestrol. J. Econ. Entomol. 81: 1288–1294.

Rufener, II. G. K., R. B. Hammond, R. L. Cooper, and S. K. St. Martin. 1987. Larval antibiosis screening technique for Mexican bean beetle resistance in soybean. Crop Sci. 27: 598–600.

Saxena, R. C. 1984. Detection of enzyme polymorphism among populations of brown plant hopper (BPH) biotypes. IRR Newsl. 9(3): 18–19.

Schmidt, D., and J. Reese, 1986. Sources of error in nutritional index studies of insects on artificial diet. J. Insect Physiol. 32: 193–198.

Schoonhoven, A. V. 1974. Resistance to thrips damage in cassava. J. Econ. Entomol. 76: 728–730.

Schukle, R. H. and L. L. Murdock. 1983. Determination of plant inhibitors by agar gel radial diffusion assay. Environ. Entomol. 12: 255–259.

Schukle, R. H., L. L. Murdock, and R. L. Gallun. 1985. Identification and partial characterization of a major gut from larvae of the Hessian fly, *Mayetiola destructor* (Say) (Diptera: Cecidomyiidae). Insecta Biochem. 15(1): 93–101.

Schuster, M. F. 1983. Screening cotton for resistance to spider mites. In *Host Plant Resistance Research Methods for Insects, Diseases, Nematodes and Spider Mites in Cotton.* Southern Cooperative Series Bulletin 280, Mississippi Agricultural and Forestry Experimental Station, pp. 54–56.

Shanks, C. H., Jr. and D. Chase. 1976. Electrical measurement of feeding by the strawberry aphid on susceptible and resistant strawberries and non-host plants. Ann. Entomol. Soc. Am. 69: 784–786.

Shaver, T. N., M. J. Lukefahr, and J. A. Garcia. 1970. Food utilization, ingestion, and growth of larvae of the bollworm and tobacco budworm on diets containing gossypol. J. Econ. Entomol. 63: 1544–1546.

Simonet, D. E., R. L. Pienkowski, D. G. Martinez, and R. D. Blackeslee. 1978. Laboratory and field evaluation of sampling techniques for the nymphal stages to the potato leafhopper on alfalfa. J. Econ. Entomol. 71: 840–842.

Singh, P. and R. F. Moore (eds.) 1985. *Handbook of Insect Rearing*, Vol. I. Elsevier, New York, 481 pp.

Smith, C. M. and N. F. Fischer. 1983. Chemical factors of an insect resistant soybean genotype affecting growth and survival of the soybean looper. Entomol. Exp. Appl. 33: 343–345.

Smith, C. M., J. L. Frazier, and W. E. Knight. 1976. Attraction of the female clover head weevil, *Hypera meles* (F.), to *Trifolium* spp. flower volatiles. J. Insect Physiol. 22: 1517–1521.

Smith, C. M., H. N. Pitre, and W. E. Knight. 1975. Evaluation of crimson clover for resistance to leaf feeding by the adult clover head weevil. Crop Sci. 15: 257–258.

Sogawa, K. 1978. Electrophoretic variations in esterase among biotypes of the brown plant hopper. IRR Newsl. 3(5): 8–9.

Sorensen, E. L. 1974. Evaluating pea aphid resistance. in D. K. Barnes (ed.). *Standard Tests to Characterize Pest Resistance in Alfalfa Varieties.* USDA NC-19: 18–19.

Sorensen, E. L. and E. Horber. 1974. Selecting alfalfa seedlings to resist the potato leafhopper. Crop Sci. 14: 85–86.

Soto, P. E. 1972. Mass rearing of the sorghum shoot fly and screening for host plant resistance under greenhouse conditions. In M. G. Jotwani and W. R. Young (eds.). *Control of Sorghum Shootfly.* Oxford Press, New Delhi, 324 pp.

Stadler, E. and F. E. Hanson. 1976. Influence of induction of host preference on chemoreception of *Manduca sexta*: behavioral and electrophysiological studies. Symp. Biol. Hung. 16: 267–273.

Starks, K. J. and R. L. Burton. 1977. Greenbugs: Determining biotypes, culturing, and screening for plant resistance. U.S. Dept. Agric. ARS Tech. Bull. 1556. 12 pp.

Stock, C. A., T. J. McLoughlin, J. A. Klein, P. Dart, and M. J. Adang. 1986. Expression of *Bacillus thuringiensis* crystal protein in *Pseudomonas cepacia* 526. Proc. 3rd Int. Symposium on the Molecular Genetics of Plant Microbe Interactions. Los Angeles, CA, p. 50. Poster 30C.

Tang, C. 1986. Continuous trapping techniques for the study of allelochemicals from higher plants. In A. R. Putnam and C. Tang (eds.). *The Science of Allelochemistry.* Wiley, New York, pp. 113–131.

Tang, C.-S. and C.-C. Young. 1982. Collection and identification of allelopathic compounds from the undisturbed root system of *Bigalta* limpograss (*Hemarthia altissima*). Plant Physiol. 69: 155–160.

Tarn, T. R. and J. B. Adams. 1982. Aphid probing and feeding, electronic monitoring, and plant breeding. In K. F. Harris and K. Maramorosch (eds.). *Pathogens. Vectors, and Plant Diseases: Approaches to Control.* Academic, New York, pp. 221–246.

Teetes, G. L. 1980. Breeding sorghums resistant to insects. In F. G. Maxwell and P. R. Jennings. (eds.). *Breeding Plants Resistant to Insects.* Wiley, New York, pp. 457–471.

Tedders, W. L. and B. W. Wood. 1987. Field studies of three species of aphids on pecan: An improved cage for collecting honeydew and glucose-equivalents contained in honeydew. J. Entomol. Sci. 22: 23–28.

Tingey, W. M. 1986. Techniques for evaluating plant resistance to insects. In J. A. Miller and T. A. Miller (eds.). *Plant Insect Interactions.* Springer, New York, pp. 251–284.

Tugwell, Jr., N. P. 1983. Methods: evaluating cotton for resistance to plant bugs. In *Host Plant Resistance Research Methods for Insects, Diseases, Nematodes and Spider Mites in Cotton.* Southern Cooperative Series Bulletin 280. Mississippi Agricultural and Forestry Experiment Station, pp. 46–53.

Vaeck, M., A. Reynaerts, H. Hofte, S. Jansens, M. De Beuckeleer, C. Dean. M. Zabeau, M. Van Montagu, and J. Leemans. 1987. Transgenic plants protected from insect attack. Nature 328: 33–37.

Velusamy, R. and E. A. Heinrichs 1986. Electronic monitoring of feeding behavior of *Nilaparvata lugens* (Homoptera: Delphacidae) on resistant and susceptible rice cultivars. Environ. Entomol. 15: 678–682.

Villani, M. G. and F. Gould. 1986. Use of radiographs for movement analysis of the corn wireworm, *Melanotus communis* (Coleoptera: Elateridae). Environ. Entomol. 15: 462–464.

Visser, J. H. 1976. The design of a low speed wind tunnel as an instrument for the study of olfactory orientation in the Colorado beetle (*Leptinotarsa decemlineata*). Entomol. Exp. Appl. 20: 275–288.

Visser, J. H. 1979. Electroantennogram responses of the Colorado potato beetle, *Leptinotarsa decemlineata* to plant volatiles. Entomol. Exp. Appl. 25: 86–97.

Visser, J. H. 1986. Host odor perception in phytophagus insects. Ann. Rev. Entomol. 31: 121–144.

Waldbauer, G. P. 1968. The consumption and utilization of food by insects. Adv. Insect Physiol. 5: 229–288.

Webster, J. A. and D. H. Smith, Jr. 1983. Developing small grains resistant to the cereal leaf beetle. USDA Tech. Bull. No. 1673. 10 pp.

Werr, W. and H. Lorz. 1986. Transient gene expression in a Graminae cell line. Mol. Gen. Genet. 20: 471–475.

White, W. H. and J. E. Irvine. 1987. Evaluation of variation in resistance to sugarcane borer (Lepidoptera: Pyralidae) in a population of sugarcane derived from tissue culture. J. Econ. Entomol. 80: 182–184.

Widstom, N. W. and R. L. Burton. 1970. Artificial infestation of corn with suspensions of corn earworm eggs. J. Econ. Entomol. 63: 443–446.

Williams, W. P. and F. M. Davis. 1985. Southwestern corn borer larval growth on corn callus and its relationship with leaf feeding resistance. Crop Sci. 25: 317–319.

Williams, W. P., P. M. Buckley, and F. M. Davis. 1985. Larval growth and behavior of the fall armyworm (Lepidoptera: Noctuidae) on callus initiated from susceptible & resistant corn hybrids. J. Econ. Entomol. 78: 951–954.

Williams, W. P., P. M. Buckley, and F. M. Davis. 1987a. Tissue culture and its use in investigations of insect resistance in maize. Agric. Ecosyst. Environ. 18: 185–190.

Williams, W. P., P. M. Buckley, and F. M. Davis. 1987b. Feeding response of corn earworm (Lepidoptera: Noctuidae) to callus and extracts of corn in the laboratory. Environ. Entomol. 16: 532–534.

Wilson, L. F. and J. L. Bean. 1959. A modified olfactometer. J. Econ. Entomol. 52: 621–624.

Wiseman, B. R. and N. W. Widstrom. 1980. Comparison of methods of infesting whorl-stage corn with fall armyworm larvae. J. Econ. Entomol. 73: 440–442.

Wiseman, B. R., F. M. Davis, and J. E. Campbell. 1980. Mechanical infestation device used in fall armyworm plant resistance programs. Fla. Entomol. 63: 425–428.

Wiseman, B. R., H. N. Pitre, S. L. Fales, and R. R. Duncan. 1986. Biological effects of developing sorghum panicles in a meridic diet on fall armyworm (Lepidoptera: Noctuidae) development. J. Econ. Entomol. 79: 1637–1640.

Woodhead, S. 1983. Distribution of chemicals in glass fibre discs used in insect bioassays. Entomol. Exp. Appl. 34: 119–120.

Wong, C., S. Ko, and S. Woo. 1983. Regeneration of rice plantlets on NaCl-stressed medium by anther culture. Bot. Bul. Acad. Sinica 24: 59–64.

7 Factors Affecting The Expression of Plant Resistance to Insects

Variation in the test insect, the plant material to be evaluated, and in the environment all affect the expression of plant resistance to insects. The ability of each variable to influence the outcome of plant resistance evaluations in the laboratory, greenhouse, or field, should be determined before drawing conclusions about relative plant resistance or susceptibility. Interactions of all three types of variables can also compound the experimental error involved in plant resistance evaluations. Therefore, it is important that these sources of error be limited as much as possible prior to the evaluation of plant material.

Environmentally induced changes in plant development have significant effects on the expression of insect resistance in plants, because plant growth is determined by existing environmental conditions. Severe environmental pertubations, such as those brought on by drought or flood conditions, cause wholesale changes in temperature and/or soil conditions. These changes then are expressed as stresses on plant metabolism and growth that, in turn, can affect the expression of plant resistance. Examples of these types of stresses are discussed at length later in this chapter. Man-made changes also affect plant growth environments, for air pollutants have been shown to have stress effects on plant growth and metabolism. Recent experimental evidence indicates that excess atmospheric sulfur dioxide alters the metabolism of bean plants and as a result increases their susceptibility to insect attack (Hughes et al. 1981, 1982). For a complete in-depth discussion of the effects of different stresses on plants in relation to insect susceptibility, readers are referred to Heinrichs (1988).

7.1 PLANT VARIABLES

7.1.1 Plant Density

The density of plant foliage affects the expression of resistance to insects. Fery and Cuthbert (1972) observed that damage to southern peas by the cowpea curculio, *Chalcodermus aeneus* Boheman, increases in proportion to changes in plant density. A 300% increase in damage was associated with a 15-fold increase in plant density. A similar relationship exists between tomato plant density and damage by larvae of the tomato fruitworm, *Heliothis zea* (Boddie) (Fery & Cuthbert 1974). Fery and Cuthbert (1973) also demonstrated that significant differences in the percentage of fruitworm damaged fruits among 22 tomato accessions are nonexistent when the data are adjusted for vine size. Harris et al. (1987) determined large differences in the number of onion fly, *Delia antiqua* (Meigen), eggs on different onion breeding lines. However, their use of covariance analysis showed that there were no differences between lines when plant size was taken into consideration.

Webster et al. (1978) determined that oats planted at intermediate and high seeding rates had higher populations of the cereal leaf beetle, *Oulema melanopus* (L.), than those planted at a lower seeding rate. Alghali (1984) determined that as rice plant spacing increases, plants produce greater numbers of tillers (stems) and that populations of the stalk-eyed fly, *Diopsis thoracica* (West), and stem damage caused by fly infestation also increase. The above examples serve to illustrate the need to standardize the spacings of plants evaluated for resistance and to express insect populations in proportion to plant biomass area.

7.1.2 Plant Height

Differences in plant height also affect the expression of resistance to insects. Tingey and Leigh (1974) detected a heigh preference of the plant bug, *Lygus hesperus* Knight, for oviposition on cotton cultivars. When unadjusted for plant height, oviposition is greatest on the tall cultivars, but when plant height differences are removed, the tall cultivars are less preferred. Smith and Robinson (1983) noted the opposite relationship between the height of rice plants and infestation by the least skipper, *Ancyloxypha numitor* (F.). Taller cultivars are less preferred, whereas short stature cultivars are heavily fed upon.

7.1.3 Plant Tissue Age

The resistance of plant tissues to insect damage varies markedly during the life of the plant and is age-specific. In some crops, plants are less resistant to insects in the early stages of development. Tolerance in some rice cultivars to the brown planthopper, *Nilaparvata lugens* Stal, is not evident in 7-day-old seedlings, but is apparent in 20-day-old plants (Velusamy et al. 1986). Plants of 'IR 36,' a rice cultivar resistant to the green leafhopper, *Nephotettix virescens* (Distant), have low levels of resistance at 10 days after planting with higher levels at 20, 40, and 60 days after planting (Rapusas & Heinrichs 1987) (Fig. 7.1). A similar trend exists in the resistance of pasture grasses to the grass aphid, *Metapolophium festucae cerealium* (Dent & Wratten 1986); resistance in the foliage of the wild tomato, *Lycopersicon hirsutum f. glabratum*, to the Colorado potato beetle, *Leptinotarsa decemlineata* (Say) (Sinden et al. 1978); in resistance of sweet corn to the corn earworm, *Heliothis zea* (Boddie), (Wann & Hills 1966); and in resistance of barley to bird-cherry aphids, *Rhopalosiphum padi* Fitch, (Leather & Dixon 1981).

Conversely, resistance in some instances is greater in younger, smaller plants.

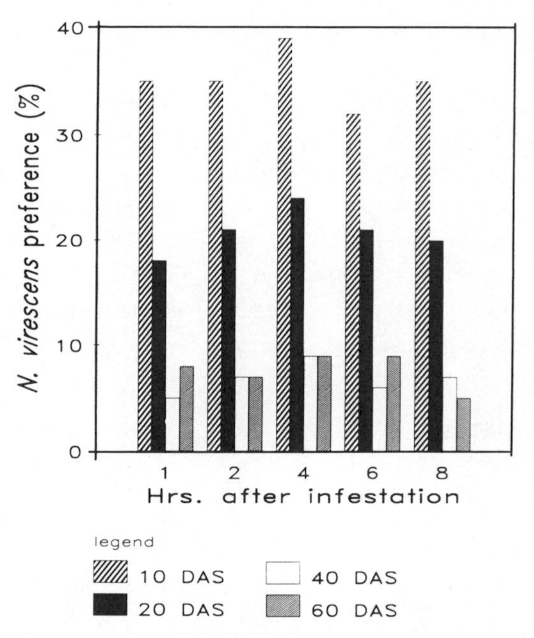

FIGURE 7.1 Effect of rice plant age 10, 20, 40, and 60 DAS (days after seeding) on the expression of resistance to the green leafhopper, *Nephotettix virescens* (Distant). (From Rapusas & Heinrichs 1987. Reprinted with permission from Environ. Entomol. 16:106–110. Copyright 1987, Entomological Society of America.)

These include resistance in sorghum to the aphid *Rhopalosiphum maidis* (Fitch), the plant hopper *Peregrinus maidis* (Ashm.) (Fisk 1978) and the migratory locust, *Locusta migratoria migratoroides* (R. & F.) (Woodhead & Bernays 1977); and resistance in maize to the European corn borer, *Ostrinia nubilalis* (Hubner) (Klun & Robinson 1969). Locust resistance in young sorghum plants is related to higher release rates of cyanide (Woodhead & Bernays 1977) (Fig. 7.2) and higher concentrations of *p*-hydroxybenzaldehyde in leaf surface waxes (Woodhead 1982). Similarly, small plants of some maize inbred lines contain more DIMBOA (see Chapters 2 and 3) than taller, more mature plants (Guthrie et al. 1986, Klun & Robinson 1969).

7.1.4 Plant Tissue Type

In the canopy of the soybean plant, younger, more succulent leaves near the plant apex are preferred for feeding by the bollworm, *Heliothis zea*, and the two-spotted spider mite, *Tetranychus urticae* (Koch) (McWilliams & Beland 1977,

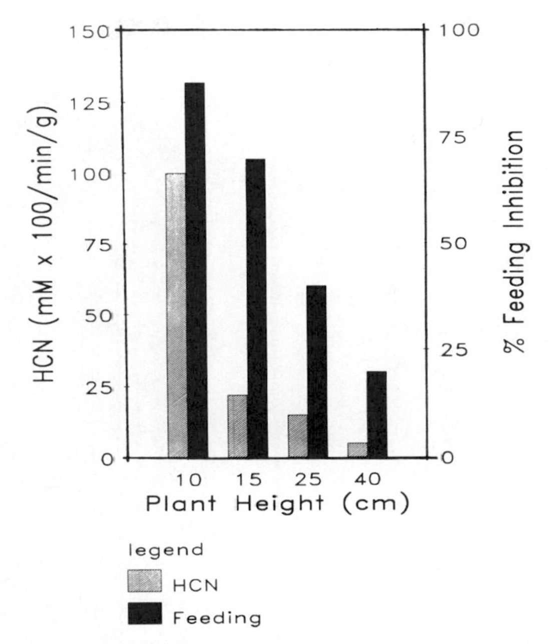

FIGURE 7.2 Relation between sorghum plant height, foliar cyanide release rate and locust feeding inhibition at varying stages of sorghum maturity (height). (Adapted from Woodhead & Bernays 1977. Reprinted by permission from Nature 270:235–236. Copyright 1977, Macmillan Magazines Ltd.)

Rodriguez et al. 1983) than lower leaves. The top two fully expanded leaves of a soybean cultivar resistant to the soybean looper, *Pseudoplusia includens* (Walker), are less resistant than other primary leaves further down the stem of the plant (Reynolds & Smith 1985) (Fig. 7.3). A similar relationship exists in the foliage of *Melilotus infesta*, a sweet clover species resistant to feeding by the sweet clover weevil, *Sitona cylindricollis* (F.) (Beland et al. 1970). In contrast, the upper leaves of the sweet pepper cultivar 'CIND' are more resistant to feeding damage from the greenhouse whitefly, *Trialeurodes vaporariorum* (Westwood), than older, more mature leaves (Laska et al. 1986).

7.1.5 Preassay Damage to Tissues

The expression of plant resistance to insects is also affected by previous exposure to various stimuli. Prior wounding by insect or mechanical means induces increased resistance of many crop plants to insect damage. Kogan and Paxton (1983) define *induced plant resistance* as the "quantitative or qualitative

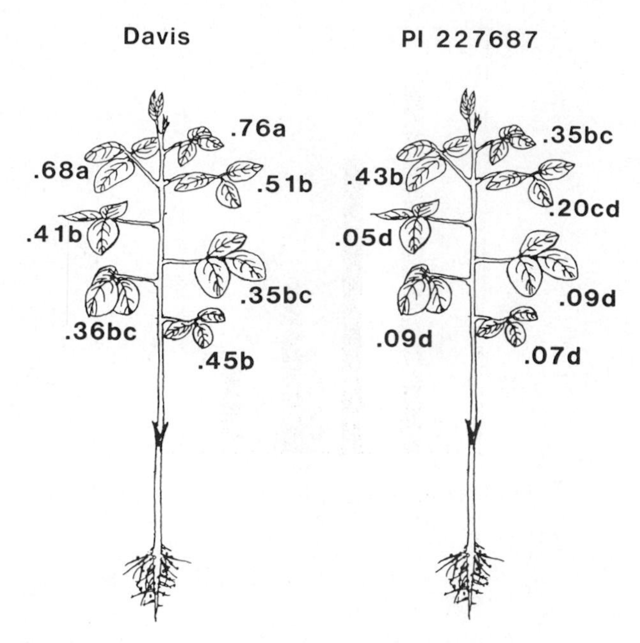

FIGURE 7.3 Susceptibility of upper trifoliates of resistant (PI227687) and susceptible ('Davis') soybean cultivars to the soybean looper, *Pseudoplusia includens* (Walker). Values are larval growth rates. (a-d, DMRT, $p < 0.01$) (From Smith 1985. Reprinted with permission from Insect Sci. Appl. 6:243–248. Copyright 1985, ICIPE Science Press.)

enhancement of a plant's defense mechanism against pests in response to extrinsic physical or chemical stimuli." Wound-induced responses in plants to mechanical damage may be part of a general defensive reaction, because some allelochemicals are similar to those developed during pathogen infection (Rhoades 1979, Edwards & Wratten 1983). Mechanical damage and insect feeding damage induce resistance in several crop plants (Table 7.1). Induction can occur in as little as 4 hours in tomato plants and may remain in effect for as long as 7 days (Table 7.1). In some tree species the induction may remain in effect for as long as 3 years.

Feeding of the bollworm on cotton plants induces the production of increased levels of phenolic compounds that cause antibiotic symptoms in bollworm larvae (Guerra 1981). Feeding of the two-spotted spider mite on cotton and soybean plants induces resistance that effectively limits further increase of mite populations (Harrison & Karban 1986, Hildebrand et al. 1986, Karban & Carey 1984, Karban 1985). Mechanical abrasion of soybean foliage

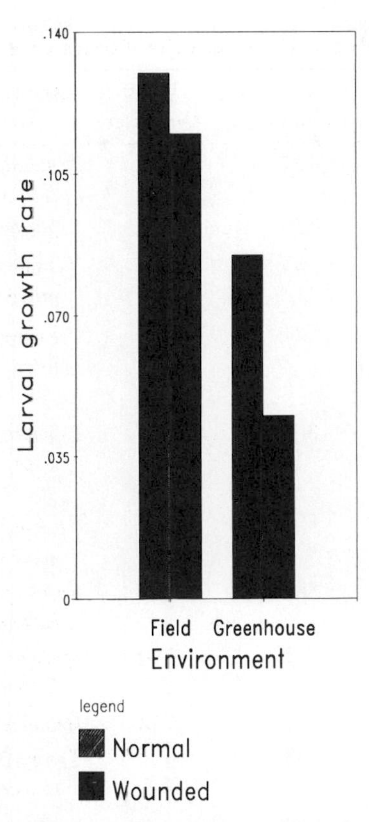

FIGURE 7.4 Induction of resistance in soybean plant introduction (PI)227687 to larvae of the soybean looper, *Pseudoplusia includens* (Walker) in field and greenhouse environments. (From Reynolds & Smith 1985. Reprinted with permission from Environ. 14:475–478. Copyright 1985, Entomological Society of America.)

also raises the level of soybean (Fig. 7.4) resistance to the soybean looper, in the foliage of an insect-resistant cultivar (Reynolds & Smith 1985).

Several species of trees also become resistant to insect damage following defoliation or mechanical damage (Table 7.1). Damage to the outer bark of apple, *Malus domestica* Borkh., trees induces resistance to the two-spotted spider mite (Ferree & Hall 1981). Foliage of paper birch trees, *Betula papyrifera* Marsh., that have been previously defoliated is more resistant to larval feeding damage of the spear-marked black moth, *Rheumaptera hastata* (L.) (Werner 1979). Similarly, the foliage of the birch trees *B. pubescens* and *B. pendula* that have been previously damaged is more resistant to larvae of the armyworm, *Spodoptera littoralis* (Walker), than undamaged foliage (Wratten et al. 1984).

TABLE 7.1 Wound-induced Resistance in Plants to Arthropods

Plant	Method of Induction[a]	Days to Induction	Days Duration	Arthropod(s) Affected	Reference(s)
Alder	I	27	44	*Malacosoma californica pluvidae*	Rhoades 1983
Apple	M	35	?	*Tetranychus urticae*	Ferree and Hall 1981
Beet	I	24	42	*Pegomya betae*	Rottger and Klinghauf 1976
Birch	I	360	720	*Lymantria dispar*	Wallner and Walton 1979
	M	2	720	*Oporinia autumnata*	Wratten et al 1984
	I	360	900	*Rheumaptera hastata*	Werner 1979
	I	$\frac{1}{4}$	60	*Spodoptera littoralis*	
Cotton	I	2	28	*Heliothis* spp.	Guerra 1981
	I, M	33	?	*Tetranychus urticae*	Karban 1985
Maple	M	3	?	Lepidoptera larvae	Baldwin and Schultz 1983
Oak	M	30–60	360	*L. dispar*	Schultz and Baldwin 1982
	I	40	120	*Phyllonorycter* spp.	West 1985
Pine	M	7	200	*Dendroctonus frontalis*	Nebeker and Hodges 1983
	I	?	360	*Neodipridon sertifer*	Thielges 1968
	I	360	360	*Panolis flammea*	Leather et al. 1987
Poplar	M	1	?	Lepidoptera larvae	Baldwin and Schultz 1983

TABLE 7.1 *(Continued)*

Soybean	M	1	30	*Pseudoplusia includens*	Reynolds and Smith 1985
Tomato	M	2	7	*S. littoralis*	Edwards and Wratten 1983
Willow	I, M	4–7	90	*Plagiodera versicolora*	Raupp and Denno 1984, Rhoades 1983
	I	35	?	*Hyphantria cunea*	
	I	11	15	*M. pluviale*	

[a]I, insect induced; M, mechanically induced.

Red oak, *Quercus rubra* L., and black oak, *Q. velutina* Lam., with a history of defoliation by the gypsy moth, *Lymantria dispar*, are more resistant to later infestations than undamaged trees (Schultz & Baldwin 1982, Wallner & Walton 1979). Feeding by the Lepidopteran leaf miner, *Phyllonorycter harrisella* (L.) also induces resistance in *Q. robur* (West 1985). Scarring the trunk of loblolly pine trees increases resistance to the southern pine beetle, *Dendroctonus frontalis* Zimmerman, for several months (Nebeker & Hodges 1983). Damage to the leaves of the willow trees *Salix alba* L. and *S. babylonica* L. imparts resistance to larvae of the imported willow leaf beetle, *Plagiodera versicolora* (Raupp & Denno 1984). Similarly, damage to the foliage of Sitka willow trees, *Salix sitchensis*, induces resistance to the fall webworm, *Hyphantria cunea* (Drury) (Rhoades 1983).

The concentration of phenolic compounds increase at the site of damaged tissues. Wounding induces two types of chemical changes involving phenolics. The first is the oxidation of pre-formed endogenous phenols, which are oxidized to produce toxic quinones, and the second is the synthesis of either mono- or diphenols, which act as defensive compounds (Rhodes & Wooltorton 1978). Increased phenolic concentrations occur in association with the induced hypersensitive responses of insect-resistant cultivars of spruce (Rohfritsch 1981, Tjia & Houston 1975) and *Solanum dulcamara* (Westphal et al. 1981) following insect feeding. Susceptible plants form cone-shaped feeding punctures, produce

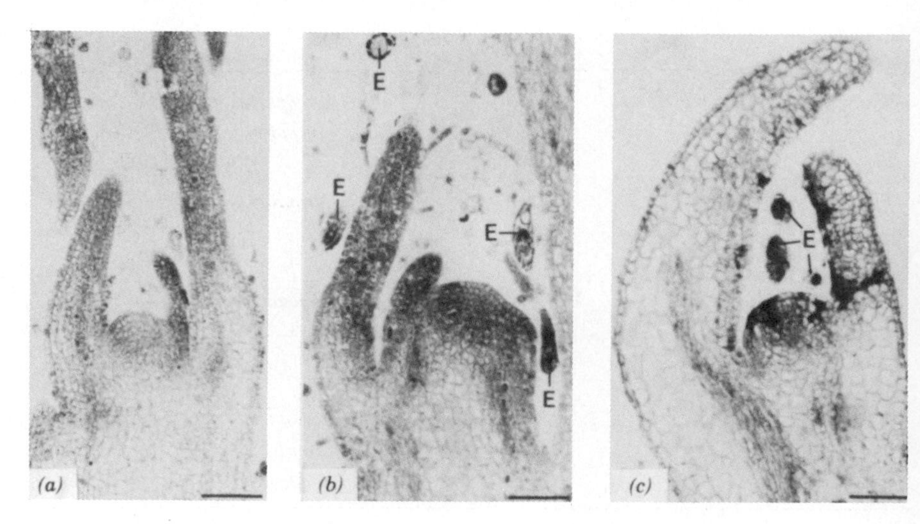

FIGURE 7.5 Cell necrosis and polyphenolic production in *Solanum dulcamara* plants following infestation by the mite, *Eriophyes cladophthirus* Nal. Longitudinal stem shoot sections of (*a*) uninfested plant; (*b*) infested susceptible plant with mites (E) causing an enlargement of the shoot base; (*c*) infested resistant plant with dark staining necrotic areas near the mites. (From Westphal et al. 1981. Reproduced by permission of the National Research Council of Canada from Can. J. Bot. 59:875–882, 1981.)

callus tissue, and form nutritive cells. Resistant *S. dulcamara* plants produce only necrotic cells near the point of attack and accumulate polyphenolic compounds after about 4 hours (Fig. 7.5). Resistance in raspberry to the raspberry cane midge, *Resseliella theobaldi* (Barnes), is due to a hypersensitive reaction to aphid feeding, which causes the formation of a wound periderm consisting of suberized and lignified cells; there is no evidence of phenol production in damaged tissues (McNicol et al. 1983).

Resistance to the bird cherry aphid in wheat is associated with the level of phenols occurring in different cultivars (Leszczynski et al. 1985). Aphid feeding induces an increase in the monophenol content of a resistant cultivar but increases polyphenol content in a susceptible cultivar. The phenol concentration of damaged tree foliage increases substantially after induction damage, and may partially explain induced resistance to larvae fed damaged birch, maple, and poplar leaves (Baldwin & Schultz 1983, Niemela et al. 1979, Wratten et al. 1984). Phenol concentrations in the foliage of pine trees attacked by the European pine sawfly, *Neodiprion sertifer* (Geoff.), the pine beauty moth, *Panolis flammea*, and the wasp, *Sirex noctilio* (F.) are higher than those of non-attacked trees (Thielges 1968, Shain & Hillis 1972, Leather et al. 1987).

7.1.6 Infection of Plant Tissues by Diseases

Plant resistance to insects may also be enhanced by the immune response of plants to invasion by infectious diseases. Ryegrass foliage with high levels of infection by the endophytic fungus, *Acremonium loliae*, is more resistant than uninfected foliage to the Argentine stem weevil, *Listronotus bonariensis* (Kuschel) (Gaynor & Hunt 1983); the bluegrass billbug, *Sphenophorus parvulus* Gyllenhal (Ahmad et al. 1986); the fall armyworm, *Spodoptera frugiperda* (J. E. Smith) (Hardy et al. 1985); the southern armyworm, *Spodoptera eridania* Cramer (Ahmad et al. 1987); and several species of sod webworms (Funk et al. 1983). Thus infections by fungal endophytes affect the expression of ryegrass resistance to insects during evaluation. The value of fungi-induced resistance in insect management is negated, however, because the fungi produce alkaloids in grasses that may cause toxicity to foraging vertebrates.

7.1.7 Evaluation of Excised and Intact Plant Tissues

Various investigators have conducted research to determine if removing tissues from plants for evaluation has an effect on the expression of resistance. Sams et al. (1975) compared the results of evaluations of *Solanum* plant material for resistance to the green peach aphid *Myzus persicae* Sulzer, using excised leaflet bioassays in the laboratory and aphid population counts on plants in the field. Correlations between the ratings of plant material in the two evaluations were highly significant, suggesting the use of excised leaf assays as a means of rapid assessment of plant material for aphid resistance. Similarly, Raina et al. (1980) found no differences in the amount of feeding of Mexican bean beetles, *Epilachna varivestis* Mulsant, on excised and intact leaves of green bean plants.

In contrast, Thomas et al. (1966) detected significantly greater survival of spotted alfalfa aphid, *Therioaphis maculata* (Buckton), nymphs on excised alfalfa trifoliates than on intact trifoliates. The differences were consistent throughout the evaluation of 33 alfalfa clones. Excision of plant tissues also effects the involvement of allelochemicals in the expression of plant resistance to insects. Leaves of cassava, normally unpalatable to the locust, *Zonocercus variegatis* (L.), are readily eaten within 1 hour after excision, owing to a drastic decline in the amount of cyanide produced by intact leaves (Bernays et al. 1977).

7.2 INSECT VARIABLES

Variables within test insects must also be compensated for, so that insect behavior reflects the response of field populations. Insect age, sex, and preassay conditioning are important variables to be worked out before proceeding with the evaluation of plant material. The peak activity period of the test insect and the level of insect infestation on test plants are also important variables to determine.

7.2.1 Insect Age

The age of the test insect is directly proportional to the amount of plant biomass it will ingest during the evaluation of plant material. Therefore, the insect age that most accurately shows differences between resistant and susceptible cultivars is the appropriate one to use. Mexican bean beetle adults feed equally on the leaves of resistant and susceptible soybean cultivars 3 days after eclosion,

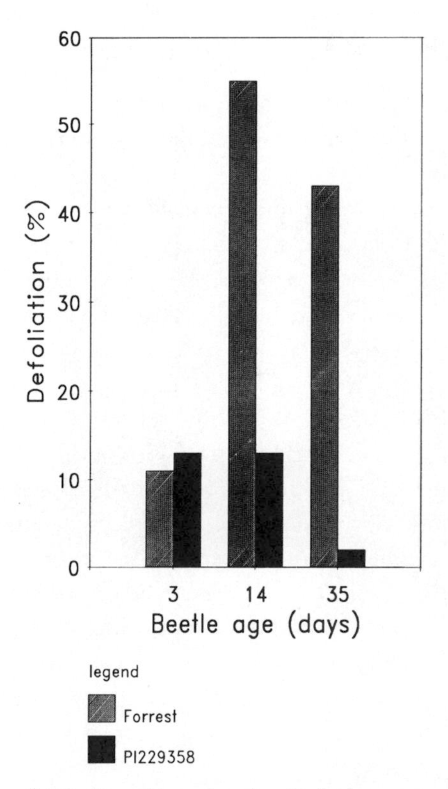

FIGURE 7.6 Effect of Mexican bean beetle, *Epilachna varvestis* (Mulsant) age on defoliation of resistant (PI229358) and susceptible ('Forrest') soybean cultivars. (From Smith et al. 1979. Adapted with permission from J. Econ. Entomol. 72:374–377. Copyright 1979, Entomological Society of America.)

but at 14 and 35 days after eclosion, differences in the amount of foliage consumed are significant (Smith et al. 1979) (Fig. 7.6). Though differences are significant at 35 days, the amount of beetle feeding is only one-half that of 14-day-old beetles.

7.2.2 Insect Sex

Sexed-based differences in the behavior of test insects also affect the outcome of plant resistance evaluations. Phytophagous female insects often consume more foliage than males, owing to the high dietary protein requirements of egg production. In addition to differential feeding, mating activity may interfere with female feeding. Schalk and Stoner (1976) determined that female

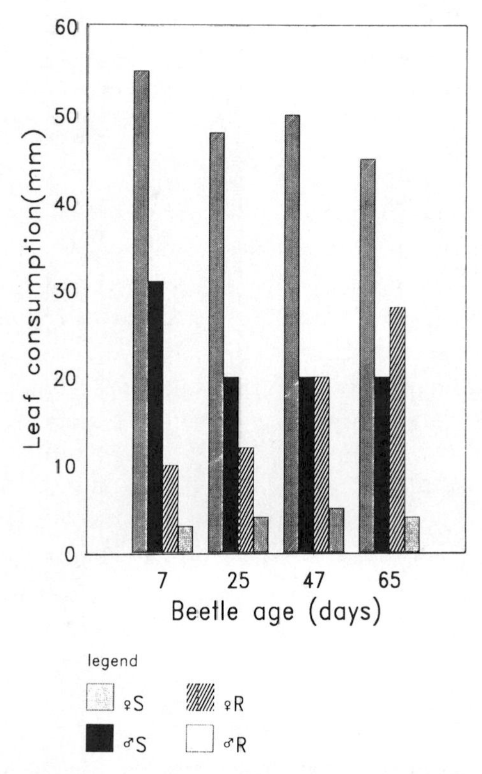

FIGURE 7.7 Consumption of resistant (R) and susceptible (S) tomato foliage by Colorado potato beetles, *Leptinotarsa decemlineata* (Say). (From Schalk & Stoner 1976. Adapted with permission from J. Am. Soc. Hortic. Sci., 101:74–76. Copyright 1976, American Society for Horticultral Science.)

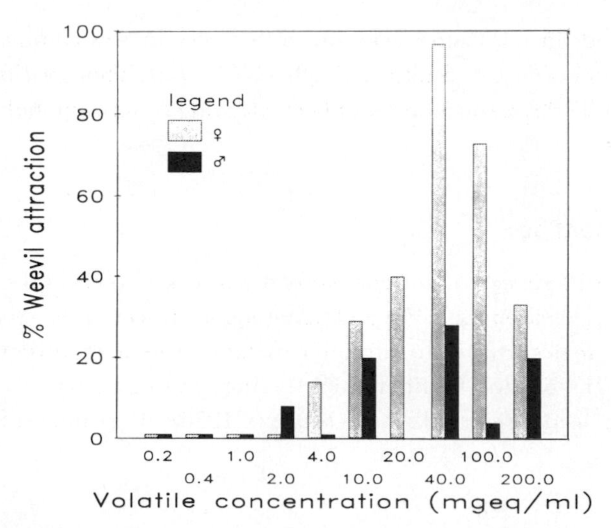

FIGURE 7.8 Attraction of clover head weevils, *Hypera meles* (F.), to flower bud volatiles of white clover. (From Smith et al. 1976. Reprinted with permission from J. Insect Physiol. Copyright 1976, Pergamon Press, Inc.)

Colorado potato beetles consumed significantly more foliage of both resistant and susceptible tomato cultivars than males (Fig. 7.7). Cook (1988) noted that female rice water weevils. *Lissorhoptrus oryzophilus* Kuschel, fed rice cultivars with varying levels of weevil resistance consumed more foliage than males. Similar results were noted by Smith et al. (1979) in evaluations of Mexican bean beetle feeding on soybean cultivars. Although sorting individuals by sex may be time-consuming, this practice adds to the accuracy of experimental results.

Sex-based responses to plant allelochemicals also exist. The attraction of clover head weevils, *Hypera meles* (F.), to *Trifolium* spp. flower bud volatiles is greater in females than in males (Smith et al. 1976) (Fig. 7.8). The attraction of ponderosa pine cone beetles, *Conopthorus ponderosae* Hopkins, to pine cone resin is also greater in females than in males (Kinzer et al. 1972).

7.2.3 Insect Activity Period

Both diel and diurnal insect activity patterns affect the accuracy of measurements of plant resistance. Warner and Richter (1974) monitored the annual abundance of adult alfalfa weevils in Oregon during day and night hours. Weevils are present in greater numbers on plants in the fall, winter, and spring during the day, but are more prevalent on plants at night during the warm summer months. Boiteau et al. (1979) detected a diel population fluctuation in the bean leaf beetle, *Cerotoma trifurcata* (Forster), on soybean plants growing in the field. Sweep net catches of beetles were lowest from 1100 to 1300 hrs and greatest after 1500 hrs. Both of the above examples stress the need to monitor insect populations from field evaluations at times when test insects are the most abundant and to avoid sampling populations during periods when insect activity patterns or habitat preferences are changing.

7.2.4 Insect Infestation Level

Determining the proper level of insect infestation on test plants is necessary in order to avoid over- or underestimation of the resistance of test plants. The starting point, and usually the minimum level of insect infestation, is one that causes economically significant damage to plants. In order to eliminate large amounts of plant material rapidly in a program of screening for resistance, heavier than normal insect infestation levels may be placed on plants. By monitoring the increase of natural populations of the pest insect in field plot border rows (Chapter 6), supplemental infestations can be placed on plants at the same time as natural populations peak. However, this procedure may mask the expression of low-level resistance in some plant material thay may have use in later stages of the breeding program.

7.2.5 Preassay Conditioning

The conditions in which insects are kept prior to testing directly affect test insect behavior during the evaluation of plant resistance. Saxena (1967) observed that the attraction of the red cotton bug, *Dysdercus koenigii* (F.), to cotton seed extracts increased in proportion to the amount of time that insects were starved prior to testing (Fig. 7.9). A similar relationship exists in the response of starved desert locusts, *Schistocerca gregaria* Forskal, to grass odors (Moorehouse 1971). Schweissing and Wilde (1979a) evaluated several small grain hosts of the greenbug, *Schizaphis graminum* (Rondani), for their ability to alter greenbug feeding preference. Barley, oats, rye, and wheat had no such effects, but the susceptible sorghum 'RS671' caused a predisposition of greenbugs to feed on sorghum. However, antixenosis to greenbugs in 'KS 30' sorghum was not influenced by prior culturing on 'RS671.' Wiseman and McMillian (1980) found that corn earworm larvae fed on the foliage of several different crop

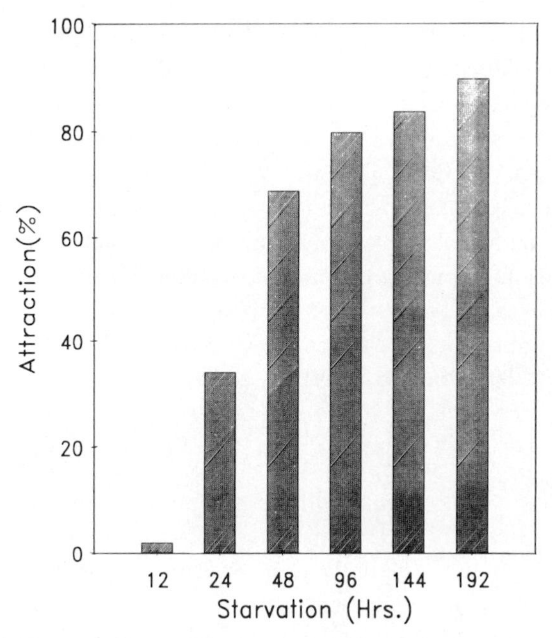

FIGURE 7.9 Response of fifth instar nymphs of the red cotton bug, *Dysdercus koenigii* (F.), to cotton seed ether extracts after various periods of starvation. (From Saxena 1967. Adapted with permission from *Olfaction and Taste II*. Copyright 1967, Pergamon Press, Inc.)

plants exhibit different feeding responses to corn silk extracts, indicating the need for standardization of the diets used to rear insects for resistance testing.

7.2.6 Insect Biotypes

Changes in the physiology of the pest insect may result in the development of resistance-breaking biotypes that alter the expression of resistance significantly. A complete discussion of biotypes is presented in Chapter 9.

7.3 ENVIRONMENTAL VARIABLES

In addition to the variation inherent in both plant and insect, the evaluation of plant resistance is also subject to variation caused by environmental effects. Such variations include those resulting from changes in lighting, temperature, relative humidity, soil nutrient conditions, and agrochemicals commonly found in contact with the crop plant. The following discussion is intended as a general guide for illustrative purposes. For a more detailed review, see Tingey and Singh (1980).

7.3.1 Light Quantity and Quality

Light quantity and quality affect the expression of plant resistance to insects. Resistance to the tobacco hornworm, *Manduca sexta* (L.), in the wild tomato, *Lycopersicon hirsutum* f. *glabratum*, is increased in plants grown under long daylengths (Kennedy et al. 1981); these produce higher quantities of 2-tridecanone, an allelochemical toxic to hornworm larvae. Resistance to the cabbage looper, *Trichoplusia ni* Hubner, in the soybean cultivar 'Niyako White' is negated by exposure of plants to continuous illumination (Khan et al. 1986). After growth of plants under a 16:8 photoperiod for 2 weeks, resistance is regained. Fluctuations in the intensity of the light received by plants also affects the expression of resistance (Table 7.2). Reduced light intensity decreases the solid stem character of wheat cultivars resistant to the wheat stem sawfly, *Cephus cinctus* (Norton) (Roberts & Tyrell 1961). Reduced light intensity also decreases exudate production by glandular trichomes on stems of insect-resistant alfalfa cultivars (Shade et al. 1975) (Fig. 7.10). Woodhead (1981) found that the lower light intensity resulting from atmospheric cloudiness reduced the production of phenolics related to insect resistance in sorghum. The expression of resistance in maize, sugar beet, and soybean is also diminished by reduced

TABLE 7.2 Decreased Plant Resistance to Insects as a Result of Reduced Light Intensity

Plant	Insect	Reference
Alfalfa	Alfalfa weevil, *Hypera postica* Gyllenhal	Shade et al. 1975
Maize	European corn borer, *Ostrinia nubilalis* (Hubner)	Manuwoto and Scriber 1985a
Sorghum	Sorghum shoot fly, *Atherigona soccata* Rond; Stem borer, *Chilo partellus* (Swinhoe)	Woodhead 1981
Sugar beet	Green peach aphid, *Myzus persicae* (Sulzer)	Lowe 1967
Wheat	Wheat stem sawfly, *Cephus cinctus* (Norton)	Roberts and Tyrell 1961

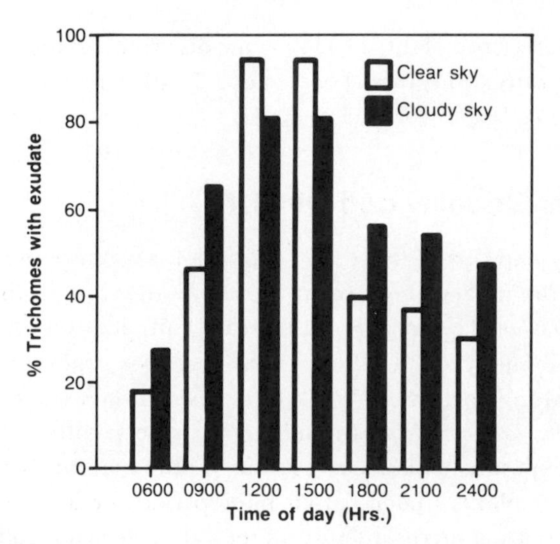

FIGURE 7.10 Effect of light intensity and time of day on exudate production by glandular trichomes of *Medicago disciformis*. (From Shade et al. 1975. Adapted with permission from *J. Econ. Entomol.* 68:399–404. Copyright 1975, Entomological Society of America.)

light intensity (Lowe 1967, Manuwoto & Scriber 1985a, Reynolds & Smith 1985).

7.3.2 Temperature

Plant resistance to insects may not be expressed at abnormally low or high temperatures. Extensive research has established that high temperature diminishes the expression of resistance in wheat to some biotypes of the Hessian fly, *Mayetiola destructor* Say (Sosa & Foster 1976, Sosa 1979, Tyler & Hatchett 1983, Ratanatham & Gallun 1986). A similar trend exists in the expression of sorghum resistance to the greenbug (Wood & Starks 1972). Alteration of the expression of insect resistance occurs at low temperatures in alfalfa. Schalk et al. (1969) found evidence of a breakdown of resistance in some alfalfa clones to the spotted alfalfa aphid at temperatures 10–15° below normal. Similar results were observed by Karner and Manglitz (1985) for alfalfa resistance to the pea aphid, *Acrythrosiphon pisum* (Harris). These examples indicate the need to monitor temperatures closely during the evaluation of plant material of all crops for insect resistance.

Insect resistance may also be affected by the temperature regime under which plants are grown. Kindler and Staples (1970a) noted that differences between susceptible alfalfa clones and clones resistant to the spotted alfalfa aphid were more apparent at fluctuating temperatures similar to natural conditions than at a constant temperature equivalent to the mean of the fluctuating regime. Similar results were recorded by van de Klashorst and Tingey (1979), who noted that ratings of potato plant material for resistance to the potato leafhopper, *Empoasca fabae* (Harris), at fluctuating temperature were more similar to field evaluation ratings than were ratings of plants grown at constant temperatures.

7.3.3 Soil Fertility

The application of varying amounts of soil nutrients to the medium in which plants are grown prior to evaluation can significantly affect the expression of resistance in these plants. Generally, increases in soil nitrogen increase plant vegetative tissue mass, supplying insects with more tissue of increased nutritional composition than plants with lower nitrogen content (Table 7.3). Increasing the amount of nitrogen fertilizer decreases fall armyworm resistance in maize (Wiseman et al. 1973), grasses (Chang et al. 1985), and peanuts (Leuck & Hammons 1974). Similar effects occur when excessive amounts of nitrogen are applied to maize resistant to the southern armyworm *Spodoptera eridania* (Cramer) (Manuwoto & Scriber 1985b), alfalfa resistant to the spotted alfalfa aphid (Kindler & Staples 1970b), and ryegrass resistant to the Argentine

TABLE 7.3 Decreased Plant Resistance to Insects as a Result of Applications of Increased Amounts of Soil Nitrogen.

Plant	Insect	Reference
Alfalfa	Spotted alfalfa aphid, *Therioaphis maculata* (Buckton)	Kindler and Staples 1970b
Maize	Fall armyworm, *Spodoptera frugiperda* (J. E. Smith)	Wiseman et al. 1973
	Southern armyworm, *Spodoptera eridania* (Cramer)	Manuwoto and Scriber 1985b
Pasture grasses	Fall armyworm	Chang et al. 1985
Peanuts	Fall armyworm	Leuck and Hammons 1974
Ryegrass	Argentine stem weevil, *Listronotus bonariensis* (Kuschel)	Gaynor and Hunt 1983

stem weevil (Gaynor & Hunt 1983). In one instance, increased amounts of nitrogen fertilizer increase the resistance of pearl millet to the fall armyworm (Leuck 1972). Increasing the amount of potassium and phosphorus fertilizer has an opposite effect to that of increases in nitrogen. The levels of spotted alfalfa aphid resistance in alfalfa (Kindler & Staples 1970b) and of greenbug resistance in sorghum (Schweissing & Wilde 1979b) increase after the addition of potassium. Resistance to the spotted alfalfa aphid in alfalfa and fall armyworm in pearl millet also increases after additional phosphorus applications (Kindler & Staples 1970b, Leuck 1972).

The above studies deal with a limited number of resistant crop cultivars and insect pests. However, the results indicate the need for standardization of the quantity and quality of soil amendments used to grow plants for insect resistance evaluations. The adoption of such protocols will allow accurate year to year comparisons of experimental results from different locations.

7.3.4 Soil Moisture

The effects of plant moisture deficiency on insect resistance have not been directly studied. However, several experiments on aphid survival demonstrate the effects of moisture stress (wilting) on plants and their aphid herbivores in general. Moisture deficiency accelerates leaf protein breakdown, causing an increase in the nitrogen content of phloem sap, and in starch hydrolysis, which increases the sucrose content of the phloem. As nitrogen and sucrose content increase, the development and survival of plant feeding aphids also increase (Kennedy & Booth 1959). At the same time, however, moisture deficiency causes the viscosity of phloem sap to increase, owing to decreased cell turgor. This increase in viscosity makes it more difficult for aphids to drink plant sap. Wearing (1972) demonstrated that at moderate wilting, brussels sprouts plants support the same number of green peach aphids as turgid plants, but during severe wilting, aphid development is restricted. As with soil nutrients, soil moisture conditions should be kept as constant as possible to obtain the most accurate measurement of insect resistance.

7.3.5 Agrochemicals

Many agrochemicals such as insecticides, herbicides, and plant growth regulators are known to have diverse effects on phytophagous arthropods (Tingey & Singh 1980, Kogan & Paxton 1983). Gall and Dogger (1967) were the first to document the effect of an agrochemical on plant resistance to insects.

Application of the herbicide 2,4-D to the wheat cultivar 'Selkirk,' which is normally susceptible to the wheat stem sawfly, results in a twofold increase in sawfly mortality. However, treatment of plants of the resistant cultivar 'Rescue' increases resistance only slightly. The effects of the growth retardant phosfon [2,4-dichlorobenzyl tributylphosphonium chloride] on two-spotted spider mite resistance in chrysanthemums were documented by Worthing (1969). Application of phosfon decreases mite survival and reproduction on the normally susceptible cultivar 'Golden Princess Anne,' but has little effect on the resistant cultivar '#4 Golden Princess Anne.'

Plant growth regulators also affect insect resistance in cotton and sorghum. The plant growth regulators (2-chloroethyl)trimethylammonium chloride (CCC) and *N,N*-dimethylpiperidinium chloride (PIX) were evaluated by Dreyer et al. (1984) for effects on greenbug resistance in sorghum. Both CCC and PIX increase resistance in the normally susceptible cultivar 'BOK 8,' but neither compound has any effect on the greenbug resistant cultivar 'G 449 GBR.' The application of PIX to cotton plants increases the content of the allelochemical gossypol, which is toxic to the tobacco budworm, *Heliothis virescens* (F.), (see Chapters 2 and 3) but has no beneficial effect on cotton resistance to budworm larvae (Mulrooney et al. 1985).

7.3.6 Relative Humidity

Relative humidity is especially important in determining plant resistance to insect pests of stored grain. Russell (1966) found that the rate of development of the lesser rice weevil, *Sitophilus oryzae* (L.) increased considerably on three sorghum cultivars normally resistant to the weevil when cultures were maintained at decrease in humidity of 20%. In a fourth cultivar, however, resistance was maintained regardless of humidity. Similar results were noted by Rogers and Mills (1974) in an evaluation of the stability of sorghum resistance to the maize weevil, *Sitophilus zeamais* (Motschulsky), at differing humidities. In this insect, however, resistance was stable in the resistant cultivar 'Double Dwarf Early Shallu' regardless of humidity, but decreased in the moderately resistant cultivar 'Redlan' as humidity increased.

7.4 CONCLUSIONS

The factors discussed in this chapter attest to the fact that plant resistance to insects is a relative phenomenon that is highly variable. As such, resistance depends on several interacting factors involving the test insect, test plant, and test environment. The differences cited here in each of these variables indicate the need to adopt standardized temperature, lighting, and humidity conditions in laboratory bioassays, so that results from experiments conducted at different times or locations can be compared. Variation in the expression of plant resistance also indicates that the condition of both plant tissues and test insects must be defined as clearly as possible, in order to minimize the variations of these organisms in plant resistance assays. In field experiments, however, the most useful insect resistance is one that remains stable over a broad range of environmental conditions. In this situation, year to year environmental variation may prove to be a useful tool for documenting the stability and operating range of various mechanisms of insect resistance.

REFERENCES

Ahmad, S., J. M. Johnson-Cicalese, W. K. Dickson, and C. R. Funk, 1986. Endophyte-enhanced resistance in perennial ryegrass to the bluegrass billbug, *Sphenophorus parvulus*. Entomol. Exp. Appl. 41: 3–10.

Ahmad, S., S. Govindarajan, J. M. Johnson-Cicalese, and C. R. Funk. 1987. Association of a fungal endopte in perennial ryegrass with antibiosis to larvae of the southern armyworm, *Spodoptera eridania*. Entomol. Exp. Appl. 45: 287–294.

Alghali, A. M. 1984. Effect of plant spacing on the infestation levels of rice by the stalk-eyed borer, *Diopsis thoracica* West (Diptera: Diopsidae). Trop. Agric. (Trinidad) 61: 74–75.

Baldwin, I. T. and J. C. Schultz. 1983. Rapid changes in tree leaf chemistry induced by damage: Evidence for communication between plants. Science 221: 277–279.

Beland, G. L., W. R. Akeson, and G. R. Manglitz. 1970. Influence of plant maturity and plant part on nitrate content of the sweet clover weevil-resistant species *Melilotus infesta*. J. Econ. Entomol. 63: 1037–1039.

Bernays, E. A., R. F. Chapman, E. M. Leather, A. R. McCaffery, and W. W. D. Modder. 1977. The relationship of *Zonocerus variegatus* (L.) (Acridoidea: Pyrgomorphidae) with cassava (*Manihot esculenta*). Bull. Entomol. Res. 67: 391–404.

Boiteau, G., J. R. Bradley, and J. W. Van Duyn. 1979. Bean leaf beetle: Diurnal population fluctuations. Environ. Entomol. 8: 615–618.

Chang, N. T., B. R. Wiseman, R. E. Lynch, and D. H. Habeck. 1985. Resistance to fall armyworm: Influence of nitrogen fertilizer application on nonpreference and antibiosis in selected grasses. J. Agric. Entomol. 2: 137–146.

Cook, C. A. 1988. Evaluation of resistance in rice to the rice water weevil *Lissorhoptrus Oryzophilus* Kuschel. M. S. thesis, Louisiana State University, Baton Rouge, 57 pp.

Dent, D. R. and S. D. Wratten. 1986. The host-plant relationships of apterous virginoparae of the grass aphid *Metopolophium festucae cerealium*. Ann. Appl. Biol. 108: 567–576.

Dreyer, D. L., B. C. Campbell, and K. C. Jones. 1984. Effect of bioregulator-treated sorghum on greenbug fecundity and feeding behaviour: implications for host-plant resistance. Phytochemistry 23: 1593–1596.

Edwards, P. J. and S. D. Wratten. 1983. Wound induced defenses in plants and their consequences for patterns of insect grazing. Oecologia 59: 88–93.

Ferree, D. C. and F. R. Hall. 1981. Influence of physical stress on photosynthesis and transpiration of apple leaves. J. Am. Soc. Hortic. Sci. 106: 348–351.

Fery, R. L. and F. P. Cuthbert, Jr. 1972. Association of plant density, cowpea curculio damage and *Choanephora* pod rot in southern peas. J. Am. Soc. Hortic. Sci. 97: 800–802.

Fery, R. L. and F. P. Cuthbert, Jr. 1973. Factors affecting evaluation of fruitworm resistance in the tomato. J. Am. Soc. Horic. Soc. 98: 457–459.

Fery, R. L. and F. P. Cuthburt, Jr. 1974. Effect of plant density on fruitworm damage in the tomato. Hortic Sic. 9: 140–141.

Fisk, J. 1978. Resistance of *Sorghum bicolor* to *Rhopalosiphum maidis* and *Peregrinus maidis* as affected by differences in the growth stage of the host. Entomol. Expl. Appl. 23: 227–236.

Funk, C. R., P. M. Halisky, M. C. Johnson, M. R. Siegel, A. V. Stewart, S. Ahmad, R. H. Hurley, and I. C. Harvey, 1983. An endophytic fungus and resistance to sod webworms: association in *Lolium perenne* L. Bio/Technology 1: 189–191.

Gall, A. and J. R. Dogger. 1967. Effect of 2, 4-D on the wheat stem sawfly. J. Econ. Entomol. 60: 75–77.

Gaynor, D. L. and W. F. Hunt. 1983. The relationship between nitrogen supply, endophytic fungus, and argentine stem weevil resistance in ryegrasses. Proc. N.Z. Grassland Assn. 44: 257–263.

Guerra, D. J. 1981. Natural and *Heliothis zea* (Boddie)-induced levels of specific phenolic compounds in *Gossypium hirsutum* (L.) M. S. thesis. University of Arkansas, Fayetteville, 85 pp.

Guthrie, W. D. Wilson, R. L., Coats, J. R., J. C. Robbins, C. T. Tseng, J. L. Jarvis, and W. A. Russell. 1986. European corn borer (Lepidoptera: Pyralidae) leaf-feeding resistance and DIMBOA content in inbred lines of dent maize grown under field versus greenhouse conditions. J. Econ. Entomol. 79: 1492–1496.

Hardy, T. N., K. Clay, and A. M. Hammond, Jr. 1985. Fall armyworm (Lepidoptera: Noctuidae): A laboratory bioassay and larval preference study for the fungal endophyte of perennial ryegrass. J. Econ. Entomol. 78: 571–575.

Harris, M. O., J. R. Miller, and O. M. B. de Ponti. 1987. Mechanisms of resistance to onion fly egg-laying. Entomol. Exp. Appl. 43: 279–286.

Harrison, S. and R. Karban. 1986. Behavioral response of spider mites (*Tetranychus urticae*) to induced resistance of cotton plants. Ecol. Entomol. 11: 181–188.

Heinrichs, E. A. (ed.) 1988. *Plant Stress–Insect Interactions*. Wiley, New York, 600 pp.

Hildebrand, D. F., J. G. Rodriquez, G. C. Brown, K. T. Luu, and C. S. Volden. 1986. Peroxidative responses of leaves in two soybean genotypes injured by twospotted spider mites (Acari: Tetranychidae). J. Econ. Entomol. 79: 1459–1465.

Hughes, P. R., J. E. Potter, and L. H. Weinstein. 1981. Effects of air pollutants on plant-insect interactions: reactions of the Mexican bean beetle to SO_2-fumigated pinto beans. Environ. Entomol. 10: 741–744.

Hughes, P. R., J. E. Potter, and L. H. Weinstein. 1982. Effects of air pollution of plant-insect interactions: increased susceptibility of greenhouse-grown soybeans to the Mexican bean beetle after exposure to SO_2. Environ. Entomol. 11: 173–176.

Karban, R. 1985. Resistance against spider mites in cotton induced by mechanical abrasion. Entomol. Exp. Appl. 37: 137–141.

Karban, R. and J. R. Carey. 1984. Induced resistance of cotton seedlings to mites. Science 225: 53–54.

Karner, M. A. and G. R. Manglitz. 1985. Effects of temperature and alfalfa cultivar on pea aphid (Homoptera: Aphididae) fecundity and feeding activity of convergent lady beetle (Coleoptera: Coccinelidae). J. Kans. Entomol. Soc. 58: 131–136.

Kennedy, J. S. and C. O. Booth, 1959. Responses of *Aphis fabae* Scop. to water shortage in hostplants in the field. Entomol. Exp. Appl. 2: 1–11.

Kennedy, G. G., R. T. Yamamoto, M. B. Dimock, W. G. Williams, and J. Bordner. 1981. Effect of daylength and light intensity on 2-tridecanone levels and resistance in *Lycopersicon hirsutum* f. *glabratum* to *Manduca sexta*. J. Chem. Ecol. 7: 707–716.

Khan, Z. R., D. M. Norris, H. S. Chiang, N. E. Weiss and S. S. Oosterwyk. 1986. Light-induced susceptibility in soybean to cabbage looper, *Trichoplusia ni* (Lepidoptera: Noctuidae). Environ. Entomol. 15: 803–808.

Kindler, S. D. and R. Staples. 1970a. The influence of fluctuating and constant temperatures, photoperiod, and soil moisture on the resistance of alfalfa to the spotted alfalfa aphid. J. Econ. Entomol. 63: 1198–1201.

Kindler, S. D. and R. Staples. 1970b. Nutrients and the reaction of two alfalfa clones to the spotted alfalfa aphid. J. Econ. Entomol. 63: 938–940.

Kinzer, H. G., B. J. Ridgill, and J. M. Reeves. 1972. Response of walking *Conophthorus ponderosae* to volatile attractants. J. Econ. Entomol. 65: 726–729.

Klun, J. A. and J. F. Robinson. 1969. Concentration of two 1, 4-benzoxazinones in dent corn at various stages of development of the plant and its relations to resistance of the host plant to the European corn borer. J. Econ. Entomol. 62: 214–220.

Kogan, M. and J. Paxton. 1983. Natural inducers of plant resistance to insects. In P. A. Hedin (ed.), *Plant Resistance to Insects*. Am. Chem. Soc. Symp. Series 208. American Chemical Society, Washington, DC, pp. 153–171.

Laska, P., J. Betlach, and M. Havrankova. 1986. Variable resistance in sweet pepper, *Capsicum annuum*, to glasshouse whitefly, *Trialeurodes vaporariorum* (Homoptera, Aleyrodidae). Acta Ent. Bohemoslov. 83: 347–353.

Leather, S. R. and A. F. G. Dixon. 1981. The effect of cereal growth stage and feeding site on the reproductive activity of the bird-cherry aphid, *Rhopalosiphum padi*. Ann. Biol. 97: 135–141.

Leather, S. R., A. D. Watt, and G. I. Forrest. 1987. Insect-induced chemical changes in young lodgepole pine (*Pinus contorta*): the effect of previous defoliation on oviposition, growth and survival of the pine beauty moth, *Panolis flammea*. Ecol. Entomol. 12: 275–281.

Leszczynski, B. 1985. Changes in phenols content and metabolism in leaves of susceptible and resistant winter wheat cultivars infested by *Rhopalosiphum padi* (L.) (Hom., Aphididae). Z. Angew. Entomol. 100: 343–348.

Leuck, D. B. 1972. Induced fall armyworm resistance in pearl millet. J. Econ. Entomol. 65: 1608–1611.

Leuck, D. B. and R. O. Hammons. 1974. Nutrients and growth media: Influence on expression of resistance to the fall armyworm in the peanut. J. Econ. Entomol. 67: 564.

Lowe, H. J. B. 1967. Interspecific differences in the biology of aphids (Homoptera: Aphididae) on leaves of *Vicia faba*. II. Growth and excretion. Entomol. Exp. Appl. 10: 413–420.

Manuwoto, S. and J. M. Scriber. 1985a. Neonate larval survival of European corn borers, *Ostrinia nubilalis*, on high and low DIMBOA genotypes of maize: effects of light intensity and degree of insect inbreeding. Agric. Ecosyst. Environ. 14: 221–236.

Manuwoto, S. and J. M. Scriber. 1985b. Differential effects of nitrogen fertilization of three corn genotypes on biomass and nitrogen utilization by the southern armyworm, *Spodoptera eridania*. Agric. Ecosyst. Environ. 14: 25–40.

McNicol, R. J., B. Williamson, P. L. Jennings, and J. A. T. Woodford. 1983. Resistance to raspberry cane midge (*Resseliella theobaldi*) and its association with wound periderm in *Rubus crataegifolius* and is red raspberry derivatives. Ann. Appl. Biol. 103: 489–495.

McWilliams, J. M. and G. L. Beland. 1977. Bollworm: Effect of soybean leaf age and pod maturity on development in the laboratory. Ann. Entomol. Soc. Am. 70: 214–216.

Moorehouse, J. E. 1971. Experimental analysis of the locomotor behavior of *Schistocerca gregaria* induced by odor. J. Insect Physiol. 17: 913–920.

Mulrooney, J. E., P. A. Hedin, W. L. Parrott, and J. N. Jenkins. 1985. Effects of PIX, a plant growth regulator, on allelochemical content of cotton and growth of tobacco budworm larvae (Lepidoptera: Noctuidae). J. Econ. Entomol. 78: 1100–1104.

Nebeker, T. E. and J. D. Hodges. 1983. Influence of forestry practices on host-susceptibility to bark beetles. Z. Angew. Entomol. 96: 194–208.

Niemela, P., E. M. Aro, and E. Haukioja. 1979. Birch leaves as a resource for herbivores. Damaged-induced increase in leaf phenols with trypsin-inhibiting effects. Rep. Kevo Subarctic Res. Stat. 15: 37–40.

Raina, A. K., P. S. Benepal, and A. Q. Sheikh. 1980. Effects of excised and intact leaf methods, leaf size, and plant age on Mexican bean beetle feeding. Entomol. Exp. Appl. 27: 303–306.

Rapusas, H. R. and E. A. Heinrichs. 1987. Plant age effect on resistance of rice 'IR36' to the green leafhopper, *Nephotettix virescens* (Distant) and rice tungro virus. Environ. Entomol. 16: 106–110.

Ratanatham, S. and R. L. Gallun. 1986. Resistance to Hessian fly (Diptera: Cecidomyiidae) in wheat as affected by temperature and larval density. Environ. Entomol. 15: 305–310.

Raupp, M. J. and R. F. Denno. 1984. The suitability of damaged willow leaves as food for the leaf beetle, *Plagiodera versicolora*. Ecol. Entomol. 9: 443–448.

Reynolds, G. W. and C. W. Smith. 1985. Effects of leaf position, leaf wounding, and plant age of two soybean genotypes on soybean looper (Lepidoptera: Noctuidae) growth. Environ. Entomol. 14: 475–478.

Rhoades, D. F. 1979. Evolution of a plant chemical defense against herbivores. In G. A. Rozenthal and D. Janzen (ed.). *Herbivores: Their Interaction with Secondary Plant Metabolites*. Academic, New York, pp. 3–54.

Rhoades, D. F. 1983. Responses of alder and willow to attack by tent caterpillars and webworms: Evidence for pheromonal sensitivity of willows. In P. A. Hedin (ed.). *Plant Resistance to Insects*. American Chemical Society Symposium Series 208. American Chemical Society, Washington, DC, pp. 56–68.

Rhodes, J. M. and L. S. C. Wooltorton. 1978. The biosynthesis of phenolic compounds in wounded plant storage tissues. In G. Kahl (ed.). *Biochemistry of Wounded Plant Tissues*. Walter de Gruyter, Berlin, pp. 243–286.

Roberts, D. W. A. and C. Tyrrell. 1961. Sawfly resistance in wheat. IV. Some effects of light intensity on resistance. Can. J. Plant Sci. 41: 457–465.

Rodriguez, J. G., D. A. Reicosky, and C. G. Patterson. 1983. Soybean and mite interactions: Effects of cultivar and plant growth stage. J. Kansas Entomol. Soc. 56: 320–326.

Rogers, R. R. and R. B. Mills. 1974. Reactions of sorghum varieties to maize weevil infestation under relative humidities. J. Econ. Entomol. 67: 692.

Rohfritsch, O. 1981. A defense mechanism of *Picea excelsa* L. against the gall former *Chormes abietis* L. (Homoptera: Adelgidae). Z. Angew. Entomol. 92: 18–26.

Rottger, V. U. and F. Klinghauf. 1976. Anderung im stoffwechsel von zuckerruben blattern durch befall mit *Pegomya bettae* Curt (Muscidae: Anthomyidae). Z. Angew. Entomol. 82: 220–227.

Russell, M. P. 1966. Effects of four sorghum varieties on the longevity of the lesser rice weevil, *Sitophilus oryzae* (L.) J. Stored Prod. Res. 2: 75–79.

Sams, D. W., F. I. Lauer, and E. B. Radcliffe. 1975. Excised leaflet test for evaluating resistance to green peach aphid in tuber-bearing *Solanum* plant material. J. Econ. Entomol. 68: 607–609.

Saxena, K. N. 1967. Some factors governing olfactory and gustatory responses in insects. In T. Hayashi (ed.). *Olfaction and Taste II.* Pergamon Press, Oxford, pp. 799–820.

Schalk, J. M. and A. K. Stoner. 1976. A bioassay differentiates resistance to the Colorado potato beetle on tomatoes. J. Am. Soc. Horic. Sci. 101: 74–76.

Schalk, J. M., S. D. Kindler, and G. D. Manglitz. 1969. Temperature and preference of the spotted alfalfa aphid for resistant and susceptible alfalfa plants. J. Econ. Entomol. 62: 1000–1003.

Schultz, J. C. and I. T. Baldwin. 1982. Oak leaf quality declines in response to defoliation by Gypsy moth larvae. Science 221: 149–151.

Schweissing, F. C. and G. Wilde. 1979a. Predisposition and nonpreference of greenbug for certain host cultivars. Environ. Entomol. 8: 1070–1072.

Schweissing, F. C. and G. Wilde. 1979b. Temperature and plant nutrient effects on resistance of seedling sorghum to the greenbug. J. Econ. Entomol. 72: 20–23.

Shade, R. E., T. E. Thompson, and W. R. Campbell. 1975. An alfalfa weevil larval resistance mechanism detected in *Medicago*. J. Econ. Entomol. 68: 399–404.

Shain, L. and W. E. Hillis. 1972. Ethylene production in *Pinus* radiata in response to *Sirex amylostereum* attack. Phytopathology 62: 1407–1409.

Sinden, S. L., J. M. Schalk, and A. K. Stoner. 1978. Effects of day length and maturity of tomato plants on tomatine content and resistance to the Colorado potato beetle. J. Am. Soc. Hortic. Sci. 103: 596–599.

Smith, C. M. 1985. Expression, mechanisms, and chemistry of resistance in soybean, *Glycine max* L. (Merr.) to the soybean looper, *Pseudoplusia includens* (Walker). Insect Sci. Appl. 6: 243–248.

Smith, C. M. and J. F. Robinson. 1983. Effect of rice cultivar height on infestation by the least skipper, *Ancyloxypha numitor* (F.) (Lepidoptera: Hesperiidae). Environ. Entomol. 12: 967–969.

Smith, C. M., J. L. Frazier, and W. E. Knight. 1976. Attraction of clover head weevil, *Hypera meles*, to flower bud volatiles of several species of *Trifolium*. J. Insect Physiol. 22: 1517–1521.

Smith, C. M., R. F. Wilson, and C. A. Brim. 1979. Feeding behavior of Mexican bean beetle on leaf extracts of resistant and susceptible soybean genotypes. J. Econ. Entomol. 72: 374–377.

Sosa, O., Jr. 1979. Hessian fly: Resistance of wheat as affected by temperature and duration of exposure. Environ. Entomol. 8: 280–281.

Sosa, O., Jr. and J. E. Foster. 1976. Temperature and the expression of resistance in wheat to the Hessian fly. Environ. Entomol. 5: 333–336.

Thielges, B. A. 1968. Altered polyphenol metabolism in the foliage of *Pinus sylvestris* associated with European pine sawfly attack. Can. J. Bot. 46: 724–725.

Thomas, J. G., E. L. Sorenson, and R. H. Painter. 1966. Attached vs. excised trifoliate for evaluation of resistance in alfalfa to the spotted alfalfa aphid. J. Econ. Entomol. 59: 444–448.

Tingey, W. M. and T. F. Leigh. 1974. Height preference of *Lygus* bugs for oviposition on caged cotton plants. Environ. Entomol. 3: 350–351.

Tingey, W. M. and S. R. Singh. 1980. Environmental factors influencing the magnitude and expression of resistance. In F. G. Maxwell and P. R. Jennings (eds.). *Breeding Plants Resistant to Insects*. Wiley, New York, pp. 89–113.

Tjia, B. and D. B. Houston. 1975. Phenolic constituents of Norway spruce resistant or susceptible to the eastern spruce gall aphid. Forest Sci. 211: 180–184.

Tyler, J. M. and J. H. Hatchett. 1983. Temperature influence on expression of resistance to Hessian fly (Diptera: Cecidomyiidae) in wheat derived from *Triticum tauschii*. J. Econ. Entomol. 76: 323–326.

van de Klashorst, G. and W. M. Tingey. 1979. Effect of seedling age, environmental temperature, and foliar total glycoalkaloids on resistance of five *Solanum* genotypes to the potato leafhopper. Environ. Entomol. 8: 690–693.

Velusamy, R., E. A. Heinrichs, and F. G. Medrano. 1986. Greenhouse techniques to identify field resistance to the brown planthopper, *Nilaparvata lugens* (Stal) (Homoptera: Delphacidae), in rice cultivars. Crop Prot. 5: 328–333.

Wallner, W. E. and G. S. Walton. 1979. Host defoliation: A possible determination of Gypsy moth population quality. Ann. Entomol. Soc. Am. 72: 62–67.

Wann, E. V. and W. A. Hills. 1966. Earworm resistance in sweet corn at two stages of ear development. Proc. Am. Soc. Hortic. Sci. 89: 491–496.

Warner, R. W. and P. O. Richter. 1974. Alfalfa weevil: Diel activity cycle of adults in Oregon. Environ. Entomol. 3: 939–945.

Wearing, C. H. 1972. Responses of *Myzus persicae* and *Brevicoryne brassicae* to leaf age and water stress in Brussels sprouts grown in pots. Entomol. Exp. Appl. 15: 61–80.

Webster, J. A., D. H. Smith, Jr., and S. H. Gage. 1978. Cereal leaf beetle (Coleoptera: Chrysomelidae): influence of seeding rate of oats on populations. Great Lakes Entomol. 11: 117–120.

Werner, R. A. 1979. Influence of host foliage on development, survival, fecundity, and oviposition of the spear-marked black moth, *Rheumaptera hastata* (Lepidoptera: Geometridae). Can. Entomol. 111: 317–322.

West, C. 1985. Factors underlying the late seasonal appearance of the lepidopterous leaf-mining guild on oak. Ecol. Entomol. 10: 111–120.

Westphal, E., R. Bronner, and M. LeRet. 1981. Changes in leaves of susceptible and resistant *Solanum dulcamara* infested by the gall mite *Eriophyes cladophthirus* (Acarina, Eriophyoidea). Can. J. Bot. 59: 875–882.

Wiseman, B. R. and W. W. McMillian. 1980. Feeding preferences of *Heliothis zea* larvae preconditioned to several host crops. J. Ga. Entomol. Soc. 15: 449–453.

Wiseman, B. R., D. B. Leuck, and W. W. McMillian. 1973. Effects of fertilizers on resistance of Antigua corn to fall armyworm and corn earworm. Fla. Entomol. 56: 1–7.

Wood, E. A., Jr. and K. J. Starks. 1972. Effect of temperature and host plant interaction on the biology of three biotypes of the greenbug. Environ. Entomol. 1: 230–234.

Woodhead, S. 1981. Environmental and biotic factors affecting the phenolic content of different cultivars of *Sorghum bicolor*. J. Chem. Ecol. 7: 1035–1047.

Woodhead, S. 1982. *p*-hydroxybenzaldehyde in the surface wax of sorghum: its importance in seedling resistance to acridids. Entomol. Exp. Appl. 31: 296–302.

Woodhead, S. and E. A. Bernays. 1977. Changes in release rates of cyanide in relation to palatability of sorghum to insects. Nature 270: 235–236.

Worthing, C. R. 1969. Use of growth retardants on chrysanthemums: Effect on pest populations. J. Sci. Food Agric. 20: 394–397.

Wratten, S. D., P. J. Edwards, and I. Dunn. 1984. Wound-induced changes in the palatability of *Betula pubescens* and *B. pendula*. Oecologia 61: 372–375.

8 Genetics and Inheritance of Plant Resistance to Insects

8.1 GENERAL

Knowledge of the genetics of insect resistance is equally as important as of the mechanism of resistance, especially if biotypes of insects are prone to develop on the crop plant under investigation. From the standpoint of breeding practicality, it is useful if resistance is associated with morphological factors of the plant leaves, stems, or fruit such as those discussed in Chapters 2 and 3. Knowledge of the genetics of resistance also enables the breeder to develop cultivars with broad bases of resistance. The genetics of insect resistance have been studied since the early twentieth century, when Harlan (1916) demonstrated that resistance to the leafblister mite, *Eriophyes gossypii* Banks, was a heritable trait in cotton.

The usual method of determining the genetic expression of resistance is to evaluate the segregating F_2 progeny from crosses between resistant and susceptible parents. *Diallel crosses* involving several resistant and susceptible parents are also used in inheritance studies. The *general combining ability* of a cultivar to transmit resistance is determined from the average resistance levels of the F_1 and F_2 plants in all crosses involving that cultivar. *Specific combining ability* is a measure of the amount of resistance transferred by a cultivar in a single cross with only one other parent. An additional standard measurement of the genetic expression of resistance involves determination of the heritability, or variation observed in the progeny of a cross that occurs owing to the additive effects of genes from resistant plants, where several different alleles contribute to resistance; to epistatic effects of alleles; or to simple dominant and recessive effects.

8.2 BREEDING METHODS USED TO DEVELOP RESISTANCE TO INSECTS

Several different methods have been utilized to breed insect resistant crop plants (Dahms 1972). The three selection methods of *mass selection, pure line selection*, and *recurrent selection* are used routinely for incorporating insect resistance into crop plants. These methods can be used in both cross- and self-pollinated plants. In self-pollinated crops, hybridization techniques such as *backcross breeding, bulk breeding*, and *pedigree breeding* are also used to incorporate insect resistance genes into agronomically desirable cultivars.

Mass selection involves selecting individual resistant plants after each generation (cycle of breeding), combining their seed, and growing this seed in the following generation as an aggregate group of plants. The objective of this breeding method is to select several sources of resistance in each of several cycles of selection. The largest improvements in resistance are usually made in the initial selection, followed by two to five additional cycles of selection. Mass selection has been effectively used to increase potato resistance to the potato leafhopper, *Empoasca fabae* (Harris) (Sanford & Ladd 1983).

Line breeding and *pure line selection* are forms of mass selection that both involve the selection of individual resistant plants that are advanced separately. In each cycle of selection, resistant selections are self-fertilized. If resistance is sought in a cross-pollinated crop, individual selections are interplanted in a later selection cycle to form a composite cross of all selected plants.

Recurrent selection is used to concentrate insect resistance genes dispersed among several different sources. In each cycle, resistant plants are selected among the progeny produced by a previous mating of resistant individuals, and the mean level of insect resistance is (it is hoped) increased. Recurrent selection allows the production of an insect-resistant cultivar with the minimum amount of inbreeding and the introduction of resistance to additional insects from different sources in later selection cycles. Recurrent selection has been used to

TABLE 8.1 Insect Resistance in Crop Plants Developed Through the Use of Recurrent Selection Breeding

Crop	Insect (s)	References (s)
Alfalfa	Alfalfa weevil, *Hypera postica* Gyllenhal	Hanson et al. 1972
	Spotted alfalfa aphid, *Therioaphis maculata* (Buckton)	Graham et al. 1965
Cotton	Boll weevil, *Anthonomous grandis* Boheman	Bird 1982
	Bollworm, *Heliothis virescens* (F.)	
	Cotton leafhopper, *Pseudatomoscelis seriatus* (Reuter)	
	Tarnished plant bug, *Lygus lineolaris* (Palisot de Beauvois)	
Maize	Corn earworm, *Heliothis zea* (Boddie)	Widstrom et al. 1982
	European corn borer, *Ostrinia nubilalis* (Hubner)	Klenke et al. 1986
	Maize weevil, *Sitophilus zeamais* Motschulsky	Widstrom 1987
Potato	Potato leafhopper, *Empoasca fabae* (Harris)	Sanford and Ladd 1987
Sweet potato	Sweet potato weevil, *Cylas formicarius elegantulus* (Summers)	Jones and Cuthbert 1973, Jones et al. 1976
Turnip	Turnip aphid, *Hyadaphis erysimi* (Kaltenbach)	Barnes and Cuthbert 1975

increase resistance to insect pests of alfalfa, cotton, maize, potato, sweet potato, and turnip (Table 8.1).

Hybridization methods allow more freedom in the selection of resistance sources, because widely divergent genotypes can be combined to obtain a higher level of genetic diversity. Hybridization consists of selection for resistance both within and between a family (population of test plants). The procedure involves selection of F_2 plants for high levels of resistance, the selection of individual resistant plants within selected F_3 families, and finally selection for resistance of entire F_4 families, with some individual plant selections for resistance. It is important to select for high levels of insect resistance and a large genetic variance in the F_2 or F_3 generation, because later cycles of selection usually involve selection for improved agronomic characteristics.

Pedigree breeding involves selection of individual plants in segregating populations on the basis of insect resistance and pedigree (Fig. 8.1). Initially a hybrid is created, and all F_1 seed are saved and replanted. The best F_2 plants are selected and planted as F_3 families. In the F_3 generation, 25–50 seeds of the resistant families are selected. In the F_4 generation, a sample of each selected resistant F_3 family (seed of 50–100 plants) is planted and selection for resistance is made within families. In the F_5 generation, samples of selected F_4 resistant families (seed of 100–500 plants) are planted, evaluated for resistance, and preliminary yield tests are conducted to eliminate resistant families with poor yields. In later generations, selections are made for families with superior resistance, yield, and other agronomic characters. The advantage of pedigree is that a great deal of susceptible plant material is eliminated early in the breeding program, allowing detailed evaluation of selected resistant plants over a period of several years.

The major disadvantages of pedigree breeding are that it is limited to use in self-pollinated crops and that only a limited number of entries can be processed owing to the extra time required for planting, harvesting, and data acquisition on each entry. Pedigree breeding has been used for increasing the levels of green leafhopper, brown plant hopper, and gall midge resistance in rice (Khush 1980). The *bulk breeding* method is also used to incorporate insect resistance into self-pollinated crops. Bulk breeding is similar to the pedigree breeding method, but selection normally does not occur until the F_5 generation.

Backcross breeding (Fig. 8.2) involves the uses of recurring backcrosses to one of the parents (recurrent parent) of a hybrid, accompanied by selection for insect resistance. The nonrecurrent parent is a source of insect resistance with a higher level of resistance than that used in the previous backcross. Backcross

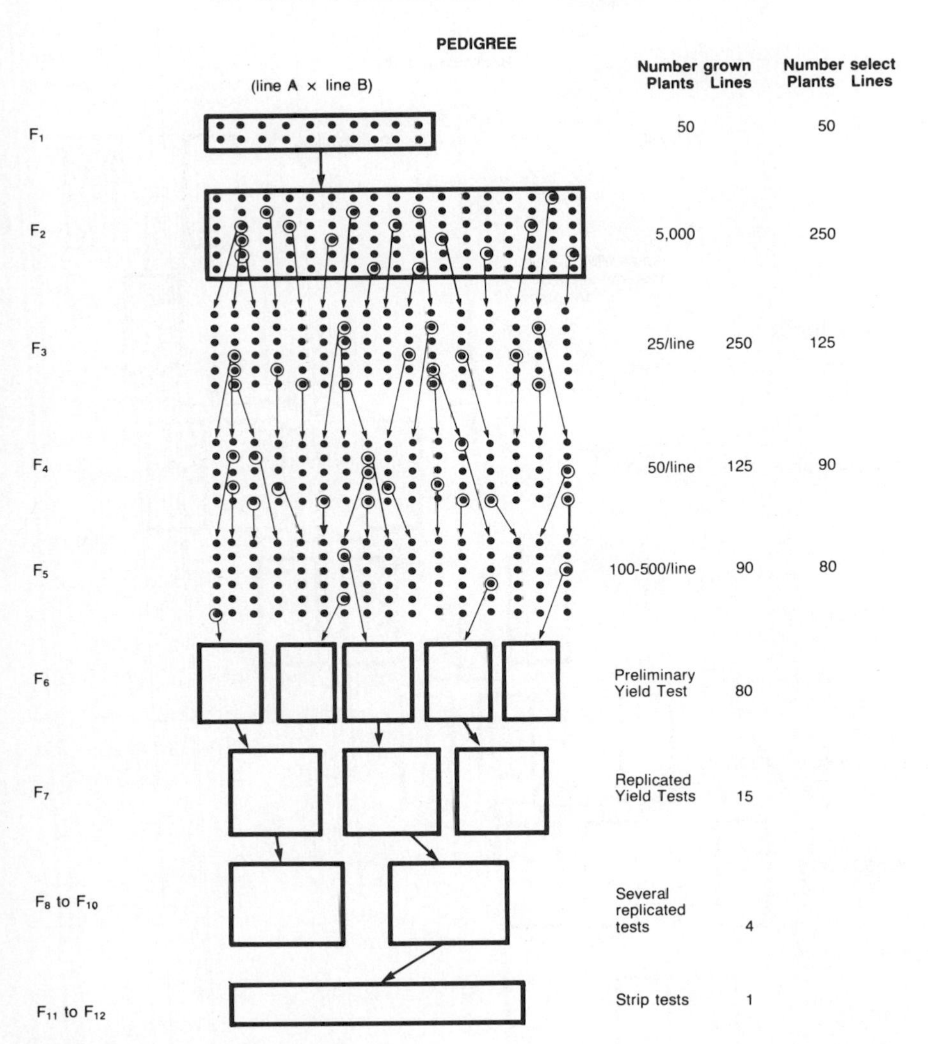

FIGURE 8.1 Schematic diagram of a pedigree breeding program for plant resistance to an insect; ⊙ indicates a selected plant.

breeding can be used as a rapid way to incorporate insect resistance into agronomically desirable cultivars that are susceptible to insects. After each cross, selections are made for agronomically desirable resistant plants. In rice, high-yielding cultivars with brown planthopper resistance have been used as recurrent parents to receive genes for resistance to other insects (Khush 1978). In soybean, backcross breeding has been used to increase the level of resistance

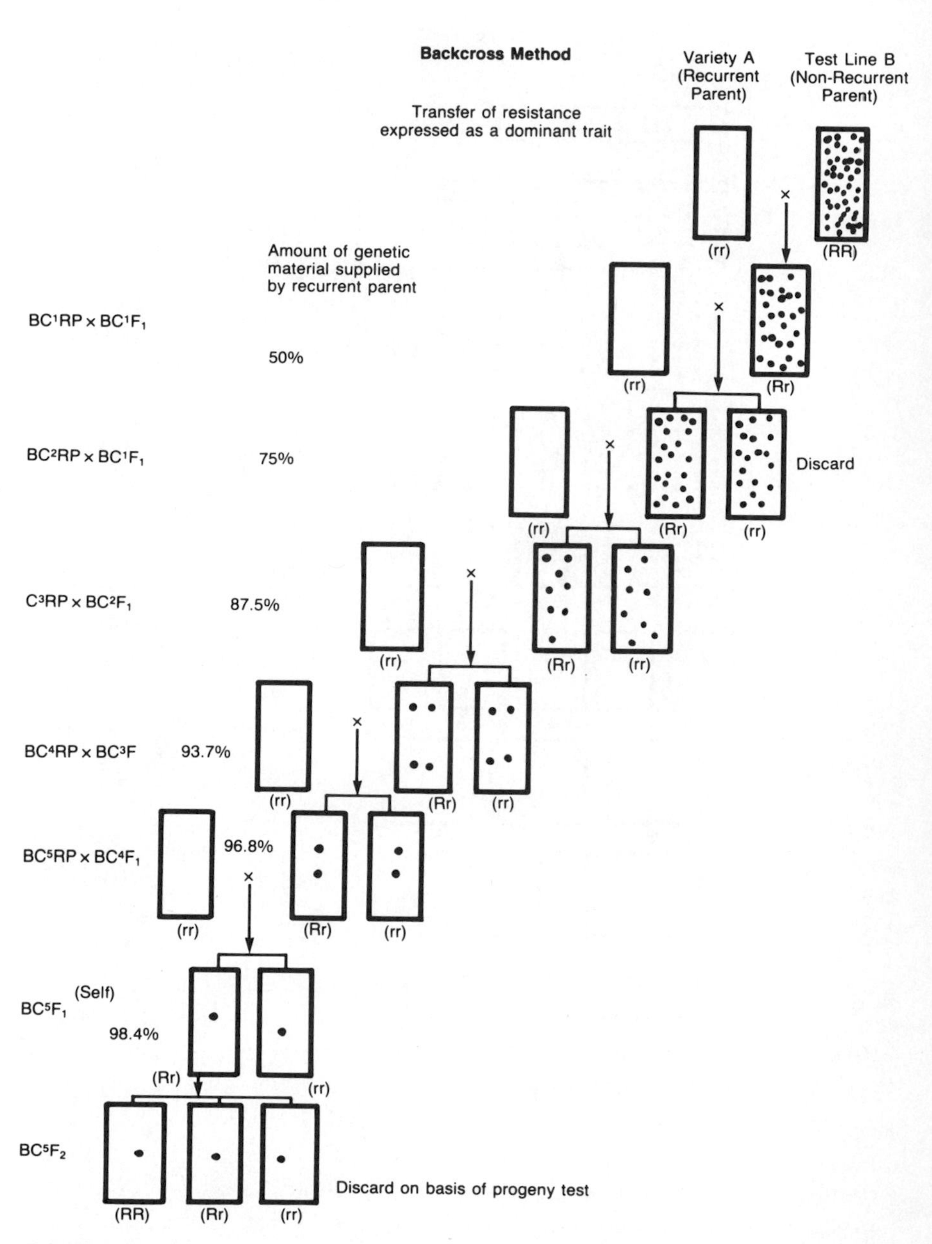

FIGURE 8.2 Schematic diagram of a backcross breeding program to transfer insect resistance from a nonrecurrent parent into a commonly grown, insect-susceptible cultivar, the recurrent parent (RP). Dots in progeny indicate the percent of genetic material (resistance genes) contributed by the resistant source.

to the Mexican bean beetle, *Epilachna varivestis* Mulsant, and corn earworm, *Heliothis zea* Boddie (Smith & Brim 1979).

8.3 GENETICS AND INHERITANCE OF RESISTANCE IN MAJOR CROPS

The genetics and inheritance of insect resistance in food, fiber, and forage crops has been extensively documented (Singh 1986). The following sections provide pertinent examples of recent progress in breeding insect-resistant crop plants.

8.3.1 Cotton

Diallel analysis indicates that additive gene effects account for approximately 90% of the total genetic variance in cotton for resistance to the tobacco budworm and gossypol gland number (Wilson & Lee 1971, Wilson & Smith 1977). Wilson & George (1979) evaluated the combining ability of resistance in cotton to seed damage by the pink bollworm, *Pectinophora gossypiella* Saunders, in a group of two cultivars and four breeding lines. In two lines, general combining ability, the average performance of a breeding line, was greatly increased for resistance to seed damage. When one of these lines was crossed with cultivars lacking extra floral nectaries, resistance was inherited because of dominant or epistatic effects (Wilson & George 1983). The gene action contributing to resistance in progeny of this cross is additive, and only a few genes condition resistance.

8.3.2 Legumes

A single dominant gene in alfalfa and sweet clover controls resistance to the pea aphid, *Acyrthosiphon pisum* (L.) (Glover & Standford 1966), and the sweetclover aphid, *Therioaphis riehmi* (Borner) (Manglitz & Gorz 1968). Resistance in alfalfa to the spotted alfalfa aphid, *Therioaphis maculata* (Buckton) is controlled by several genes (Glover & Melton 1966), indicating that resistance is quantitative.

Soper et al. (1984) performed a diallel analysis of three synthetic alfalfa populations to determine combining ability effects for potato leafhopper resistance. Significant effects for both general and specific combining ability indicate that improvements can be made in each population for leafhopper resistance. The relationship of the combining ability and inheritance of alfalfa leaf and stem trichomes for potato leafhopper, *Empoasca fabae* (Harris),

TABLE 8.2 Resistance Genes in Insect-resistant Plant Material

Crop	Insect	Gene(s)[a] (Action)	Source	Reference(s)
Barley	Greenbug	grb2	Dobaku	Cutis et al. 1960
		Grbl	Omugi	Joppa et al. 1980
Cucumber	Spotted cucumber beetle	Bi	Hawkesbury	DaCosta and Jones 1971
Lettuce	Leaf aphid	La1	*Lactuca virosa*	Eenink et al. 1982b
Maize	European corn borer (1st brood)	5–6/(A)	B52	Scott and Guthrie 1967
	(2nd brood)	7/(A)	B52	Chiang and Hudon 1973
	Western corn rootworm	r	—	Sifuentes and Painter 1964
Pearl millet	Fall armworm, corn earworm	tr	Tift 235	Burton et al. 1977
Potato	Green peach aphid	2/(Q)	*Solanum berthaultii*	Mehlenbacher et al. 1984
Rice	Brown plant hopper	Bph1, bph2	Mudgo, ASD 7	Athwal and Pathak 1971
		Bph3, bph4	Babawee, Rathu Heenati	Lakashminarayana and Khush 1977
		bph2, Bph3	Ptb21	Ikeda and Kaneda 1981
		3-4/(Q)	Utri Rajapan	Panda and Heinrichs 1983
	Gall midge	Pd	WC1263	Sastry and Prakasa Rao 1973, Satyanarayanaiah and Reddi 1972
	Green leafhopper	G1h1, G1h2, G1h3	Pankhari 203, ASD 7, IR 8	Athwal and Pathak 1971
		glh4, Glh5	Ptb 8, ASD 8	Siwi and Khush 1977
		Glh6, Glh7	Maddai Karuppan, TAPL 796	Rezual Kamin and Pathak 1982

	White-backed plant hopper	Wbph1, Wbph2	N22, ARC 10239	Sidhu et al. 1979, Angeles et al. 1981
		Wbph3, wbph4	ADR 52, Podiw-A8	Hernandez and Khush 1981
	Zigzag leafhopper	Z1h1, Z1h2, Z1h3	Rathu Heenati, Ptb 21, Ptb 33	Angeles et al. 1986
Rye	Greenbug	Gb	Caribou	Livers and Harvey 1969
Sorghum	Greenbug	Gb	Shallu Grain	Weibel et al. 1972
	Chilo partellus	(Q)	IS18327, 18489	Pathak and Olela 1983
Soybean	Mexican bean beetle	2-3 (Q)	PI229358	Sisson et al. 1976
	Cabbage looper			Leudders and Dickerson 1977
	Soybean looper			Kilen et al. 1977
Sweet clover	Sweet clover aphid	Sca	G337, G232	Manglitz and Gorz 1968
Wheat	Greenbug	gb	DS28A	Painter and Peters 1968
	Hessian fly	H1H1H2H2, H3H3, H4H4 H5H5, H6H6 + 3, H7H8	Dawson, Monon, Java, Riberio, Knox 62 Seneca	Gallun and Patterson 1977
		H9H9, H10H10	Ella, *T. turgidum*	Stebins et al. 1982, 1983
		H11H11	*T. turgidum*	Stebbins et al. 1982
		H12	Lusco	Maas et al. 1987

[a]Capital letters denote dominance; lowercase letters denote recessive traits; A = additive, Q = quantitative.

resistance was determined by Elden et al. (1986) in clones of resistant and susceptible alfalfas. Diallel analyses detected highly significant combining ability effects for the resistant clones, suggesting that trichome densities can be increased by selective breeding.

Elden and Elgin (1987) increased the level of potato leafhopper resistance in five alfalfa populations, using both recurrent selection and individual plant selection. The level of resistance in red clover to the pea aphid, *Acyrthosiphon pisum* (L.), and the yellow clover aphid, *Therioaphis trifolii* (Monell), has also been increased by recurrent selection (Gorz et al. 1979). Five cycles of selection for yellow clover aphid resistance and three cycles of selection for pea aphid resistance were employed to develop the synthetic resistant cultivar 'N-2.'

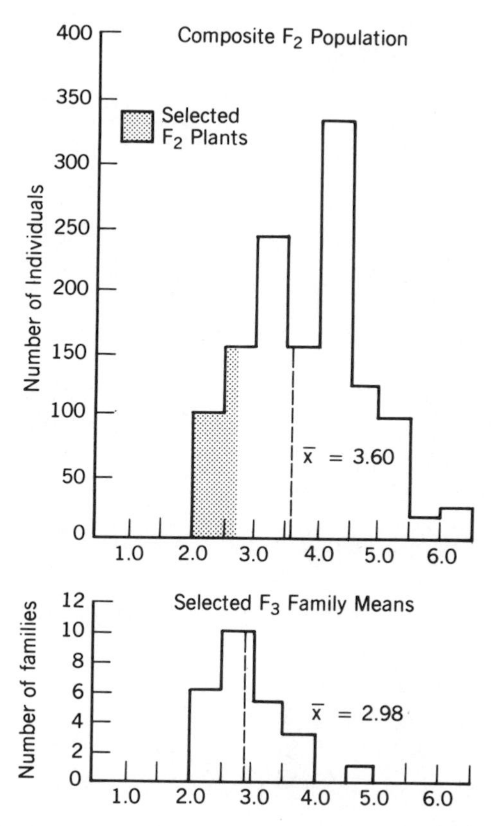

FIGURE 8.3 Enhancement in the level of resistance to pea leaf weevil, *Sitona lineatus* (L.), in F_3 plant families of Austrian winter peas selected from resistant F_2 plants. (From Nouri-Ghanbalani et al. 1978. Reprinted with the permission of the Crop Science Society of America, Inc.)

Resistance in soybean to defoliating insects is related to multiple gene effects. Heritability estimates reported for resistance to the cabbage looper, *Trichoplusia ni* (Hubner) (Luedders & Dickerson 1977), and the Mexican bean beetle (Sisson et al. 1976) suggest that resistance is quantitatively inherited. F_2 plants from a cross involving parents resistant and susceptible to the soybean looper, *Pseudoplusia includens* (Walker), also exhibit partial dominance or a quantitative inheritance action (Kilen et al. 1977).

Resistance to the pea leaf weevil, *Sitona lineatus* (L.), in Austrian winter peas is increased by 15% after selecting resistant F_2 plants from a cross between resistant and susceptible parents (Nouri-Ghanbalani et al. 1978) (Fig. 8.3). Diallel analyses of eight different genotypes indicated that differences in resistance were due to specific combining ability and that resistance could not be expressed in commercial pea cultivars. Evaluation of a large group of plant introductions subsequently identified plant materials with sufficient general combining ability to allow resistance to be transferred to commercial cultivars (Auld et al. 1980).

Resistance in lima bean, *Phaseolus lunatus*, to the leafhopper, *Empoasca kraemeri* Ross and Moore, is due to a quantitative effect of several genes and is inherited as a recessive trait (Lyman & Cardona 1982). Both additive and dominant gene effects are responsible for *E. kraemeri* resistance in cultivars of the common bean *Phaseolus vulgaris* (Kornegay & Temple 1986). There is also evidence for *transgressive segregation* (levels of resistance greater than that of the resistant parent) in some progenies from crosses between resistant and susceptible bean cultivars.

8.3.3 Fruits and Vegetables

Resistance in fruit to several species of aphids is controlled by the action of single dominant genes. These include resistance in apple to the rosy apple aphid, *Dysaphis plantaginea* (Alston & Briggs 1970), and the rosy leaf curling aphid, *Seppaphis devecta* (Alston & Briggs 1968); and in raspberry to the raspberry aphid, *Amphorophora rubi*, (Daubeny 1966) (The interaction of each of these genes to aphid biotype development is discussed in Chapter 9.)

Scott (1977) successively selfed carrot cultivars for three generations to increase the level of resistance to a complex of the western plant bug, *Lygus hesperus* Knight, and *Lygus elisus* van Duzee. *Lygus* mortality increased from 24% on plants of the S_1 generation to 85% on plants of the S_3 generation.

The genetics and inheritance of cowpea, *Vigna unguiculata*, resistance have been investigated for several insects. Bata et al. (1987) studied segregating F_2

and F_3 progeny from crosses between resistant and susceptible cowpeas and determined that resistance to the cowpea aphid, *Aphis craccivora* Koch, is inherited as a monogenic dominant trait. Conversely, resistance to bruchids infesting cowpeas is conferred in a complex inheritance pattern by a combination of major and minor genes expressed as a recessive trait. Segregation occurs among resistant breeding lines as late as the F_5 generation (Redden et al. 1983, 1984). Resistance to the cowpea weevil, *Callosobruchus maculatus* F., is controlled by both additive and dominance effects (Fatunla & Badaru 1983). Cowpea resistance to the cowpea curculio *Chalcodermus aeneus* Boheman, is also additive, and probably controlled by one pair of genes (Fery & Cuthbert 1975).

Resistance to the two-spotted spider mite, *Tetranychus urticae* Koch, in cucumber was increased by crossing several moderately resistant cultivars (de Ponti 1979), indicating that resistance is inherited polygenically. The inheritance of cucurbitacin, an insect feeding deterrent in cucumbers, is controlled by a single gene. However, resistance to the spotted cucumber beetle, *Diabrotica undecimpunctata howardi* Barber, is controlled by two or three gene pairs. Factors other than low cucurbitacin content in seedlings appear to condition the expression of resistance (Sharma & Hall 1971).

Resistance in lettuce to the leaf aphid, *Nasonovia ribis nigri*, was transferred from *Lactuca virosa* to *Lactuca sativa* by interspecific crossing (Eenink et al. 1982a). Resistance is monogenic and inherited as a dominant trait (Eenink et al. 1982b).

Cuthbert and Jones (1972) determined that recurrent selection increases insect resistance in sweet potatoes. In a random interbreeding population of sweet potato genotypes, four cycles of selection increased the incidence of resistance to a complex of soil insect consisting of the grub *Plectris aliena* Chapin; the southern potato wireworm, *Conoderus falli* Lane; the banded cucumber beetle, *Diabrotica balteata* LeConte; the spotted cucumber beetle, *Diabrotica undecimpunctata howardi* Barber; the elongate flea beetle, *Systena elongata* (F.); and *Systena frontalis* (F.).

Progenies from crosses between the cultivated potato, *Solanum tuberosum*, and four *Solanum* species with varying densities of glandular trichomes exhibit heritabilities for resistance to the green peach aphid, *Myzus persicae* (Sulzer), ranging from 50 to 60%. Resistance is expressed as a partially dominant trait (Sams et al. 1976). Glandular trichome-mediated resistance to the green peach aphid in *Solanum tarijense* and *S. berthaultii* is controlled by a single dominant gene, but in *S. phureja* × *S. berthaulti* crosses, two genes are involved in the expression of resistance (Gibson 1979). Mehlenbacher (1983, 1984) conducted heritability studies on the density of two types of glandular trichomes on *S. berthaultii* foliage in crosses involving *S. berthaultii* and *S. tubersosum*. The density of both type A trichomes (short trichomes with a four-lobed tip) and type B trichomes (long, simple trichomes with a single exudate droplet at the tip) are not highly heritable. However, the droplet size of the type B trichomes is highly heritable. None of the three traits alone is highly correlated to green peach aphid resistance, but some F_2 and F_3 progeny contain high levels of aphid resistance. Resistance is thought to be due to a complex interaction of both trichome density and droplet size. For this reason, resistance is viewed from a breeding perspective as a quantitatively inherited trait.

Segregation for resistance to the Colorado potato beetle, *Leptinotarsa decemlineata* Say, also occurs in the progeny of crosses involving *S. tuberosum* and *S. berthaultii* (Wright 1985). Recurrent selection techniques have been used to improve potato resistance to the potato leafhopper. Seven cycles of selection resulted in a 68% reduction in leafhopper damage in potato populations selected for resistance (Sanford & Ladd 1987). About one-half of this resistance is inherited in progeny from crosses between leafhopper susceptible clones and the resistant population.

Theurer et al. (1982) used mass selection techniques to isolate and identify sugar beet plant material with resistance to the sugarbeet root maggot, *Tetanops myopaeformis* Roder. The rate of change in resistance remained unchanged for five cycles of selection, indicating that higher levels of resistance remain to be obtained.

Inheritance of the phytomelanin (achene) layer in sunflower species resistant to the sunflower moth, *Homoeosoma electellum* (Hulst.), was studied by Johnson and Beard (1977). Segregation in progenies between a cross of a susceptible cultivar and a resistant *Helianthus* species indicates that resistance is inherited as a dominant trait.

8.3.4 Maize

Several genetic sources of maize have been identified that possess different numbers of genes for resistance to the European corn borer, *Ostrinia nubilalis* (Hubner). Borer resistance is conditioned by several genes (Scott & Guthrie 1967, Chiang & Hudson 1973). Jennings et al. (1974) used a diallel analysis of two sets of maize inbred lines to study the inheritance of resistance to first and second brood (generation) European corn borers. Different genes condition resistance to both first and second broods, but some genes condition resistance to both broods. Reciprocal translocation studies also demonstrate that only two or three of the 12 chromosome arms in maize are similar for both first and second brood resistance (Onukogu et al. 1978).

Recurrent selection has been used to increase the levels of resistance in maize to both first and second brood populations of the European corn borer (Klenke et al. 1986, Russell et al. 1979). Borer damage ratings decrease significantly for both broods after only four cycles of selection. Variability for borer resistance in each plant population also decreases with each successive cycle of selection.

Resistance in maize to the corn earworm, *Heliothis zea* Boddie, is also quantitatively inherited (Widstrom & Hamm 1969). Widstrom (1972) and Widstrom and McMillian (1973) determined that general combining ability is more important in earworm resistance than the dominance or epistatic effects represented by specific combining ability. Recurrent selection has been used to increase the level of resistance in maize to the corn earworm (Widstrom et al. 1982), the fall armyworm, and the maize weevil, *Sitophilus zeamais* Motschulsky, in maize breeding populations (Widstrom 1988). Testcrossing, a procedure to identify resistance levels in maize breeding lines in a recurrent selection program, has also been used to increase the level of earworm resistance in maize populations. Genetic variability for fall armyworm, *Spodoptera frugiperda* (J. E. Smith), resistance in maize is also additive (Widstrom et al. 1972; Widstrom 1976).

Backcross breeding has not generally proven successful to maize breeders, for reciprocal translocation studies have shown that at least 12 genes are involved in European corn borer resistance (Scott et al. 1966, Onukogu et al. 1978). However, Gracen (1988) used backcrossing to develop several maize inbreds with European corn borer resistance.

8.3.5 Rice

Rice plant materials are developed for insect resistance evaluation using single crosses to create F_1 hybrid plants. These plants are used to make topcrosses (an F_1 insect resistant hybrid × a commonly grown cultivar) or double crosses (F_1 hybrid × F_1 hybrid). Seeds from these crosses are planted and plants evaluated for both insect and disease resistance. F_2 plants possessing both traits (about 25% of the original F_1 population) are then placed in pedigree nurseries that are evaluated separately for insect and disease resistance. These plants are evaluated as F_3 and F_4 families for resistance and as F_5 and F_6 families for agronomic desirability and yield (Fig. 8.4) (Khush 1980).

Genes exist in rice for resistance to the brown plant hopper, *Nilaparvata lugens* Stal, the green rice leafhopper, *Nephotettix virescens* Distant, the white-backed

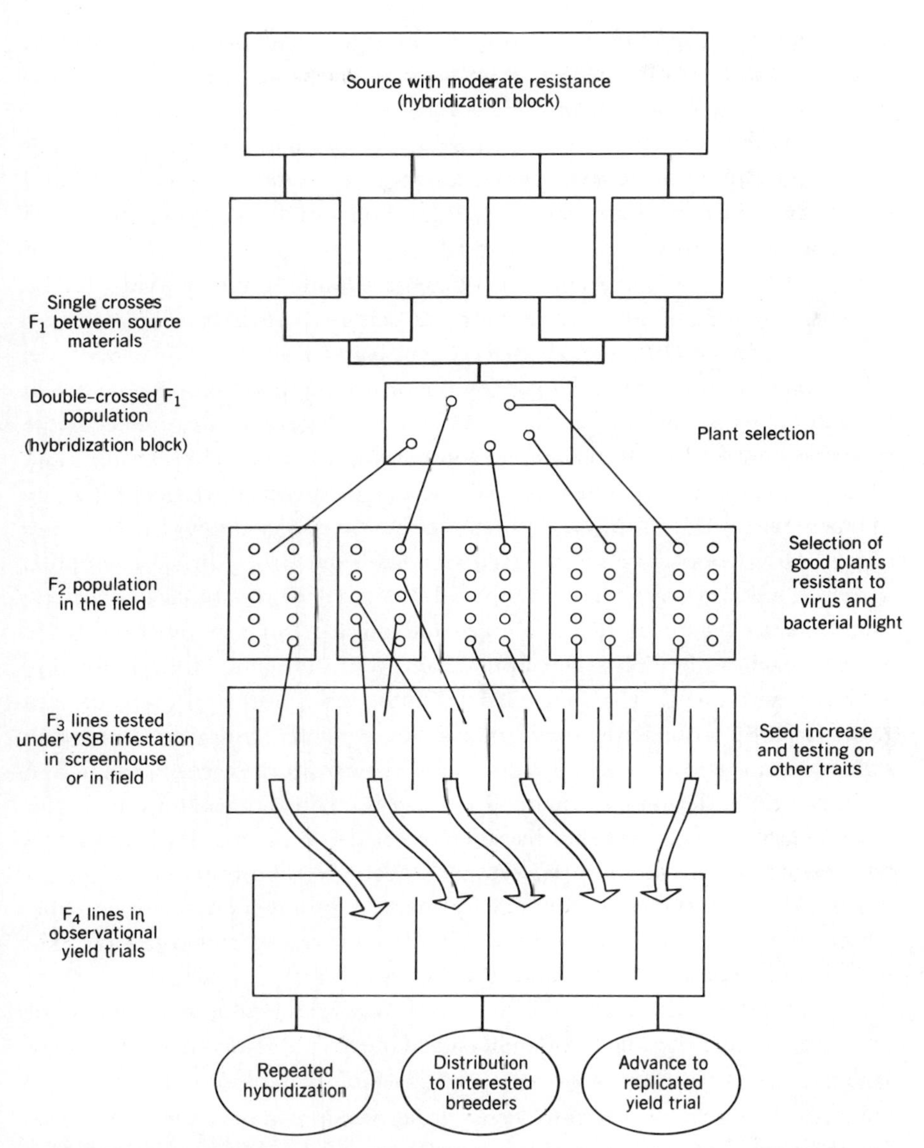

FIGURE 8.4 Hybridization scheme of crosses made in rice for multiple pest resistance to the yellow stem borer, *Tryphoryza incertulas* Walker, virus, and bacterial blight (from Khush 1980).

plant hopper, *Sogatella furcifera* Horvath; the zigzag leafhopper, *Reclia dorsalis* Motschulsky; the rice gall midge, *Orseolia oryzae* (Wood-Mason); and the yellow stem borer, *Tryporyza incertulas* (Walker).

Four genes control resistance in rice to the brown plant hopper. Two genes (Bph 1 and Bph 3) are inherited as dominant traits, and two genes (bph 2 and bph 4) are inherited as recessive traits. The Bph 1 and bph 2 genes were first noted in the cultivars 'Mudgo' and 'ASD 7,' respectively, by Athwal and Pathak (1971). The Bph 3 and bph 4 genes are found in the cultivars 'Rathu Heenati' and 'Babawee,' respectively (Lakshminarayana & Khush 1977). Sidhu & Khush (1978) evaluated 20 cultivars of rice to determine the inheritance of resistance to damage by the brown plant hopper. Resistance in some cultivars is expressed as the dominant Bph 3 gene and in others as the recessive bph 4 gene. Resistance in 'Sina Sivapu,' 'Sudu Hondawala,' and 'Ptb 33' is controlled by one recessive and one dominant gene that segregate independently. One of the two genes in 'Sinna Sivapu' is thought to be either Bph 3 or bph 4. Bph 1 and bph 2 segregate independently of Bph 3 and bph 4, and each of the two gene pairs (Bph 3 + bph 4 and Bph 1 + bph 2) are closely linked (Ikeda & Kaneda 1981). The genes involved in 'Sinna Sivapu' resistance are also likely to be bph 2 and Bph 3, for 'Ptb,' a progeny of Sinna Sivapu, exhibits resistance that is controlled by both bph 2 and Bph 3. Ikeda and Kaneda (1981) also used trisomic analyses to demonstrate that the Bph 3 and bph 4 genes are located on chromosome 7 in rice. More recently, Ikeda and Kaneda (1986) developed a method to determine the gene actions in rice for brown plant hopper resistance based on the reaction to biotypes 2 and 3 and testcrossing to cultivars of a known genotype. Several cultivars were determined to have resistance conditioned by unknown dominant or recessive genes. Three different dominant genes for resistance to the zigzag leafhopper also exist in 'Rathu Heenati,' 'Ptb 21,' and 'Ptb 33' (Angeles et al. 1986).

Quantitatively inherited resistance to brown plant hopper mediated by several genes was identified by Panda and Heinrichs (1983) in the cultivar 'Utri Rajapan.' Field resistance or tolerance in 'Utri Rajapan' and 'Triveni' is governed by two independently segregating recessive genes (Velusamy et al. 1987). Seven genes have been determined to control the expression of resistance in rice to the green leafhopper (Chelliah 1986). Genes Glh 1, 2, 3, 5, 6 and 7 are all inherited as dominant traits, but only glh 4 is inherited as recessive trait. Glh 1, 2, and 3 were identified in the cultivars 'Pankari 203,' 'ASD 7,' and 'IR 8,' respectively (Athwal and Pathak 1971), and glh 4 and Glh 5 were identified in Ptb 18 and ASD8, respectively (Siwi & Khush 1977). Glh 6 and Glh 7 were identified in the cultivars 'TAPL 796' and 'Maddai Karuppan,' respectively, by Rezaul Kamin and Pathak (1982). Two genes with dominant effects for green leafhopper resistance exist in the cultivar 'IR 42' (Takita & Habibuddin 1986).

Resistance to the gall midge is related to the effects of from one to three dominant genes (Sastry & Prakasa Rao 1973, Satyanarayanariah & Reddi 1972). Chaudhary et al. (1986) investigated the genetics of rice resistance to the Raipur, India gall midge biotype. Segregation ratios indicate that each of five known resistant cultivars has one dominant resistant gene. Allele tests with the progeny of crosses between the resistant sources indicate that three cultivars have the same gene for resistance (Gm-1). The other two cultivars both have the same dominant gene for resistance (Gm-2) that is nonallelic to Gm-1. The

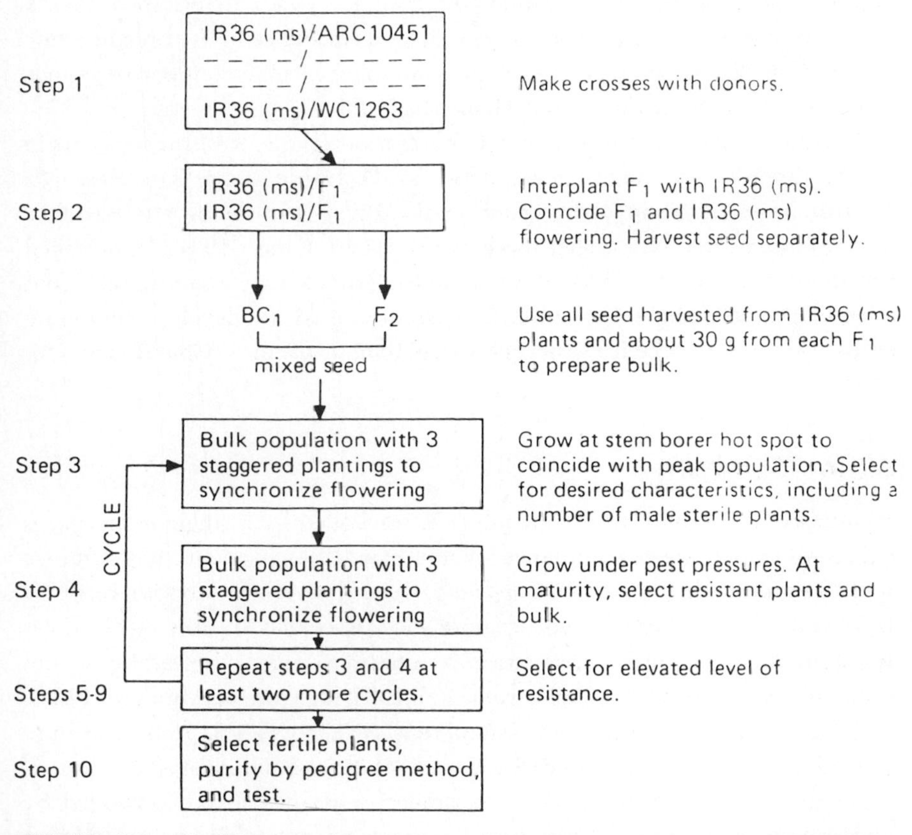

FIGURE 8.5 Modified breeding strategy for breeding resistance in rice to the yellow stem borer, *Tryphoryza incertulas* Walker, using recurrent selection, pedigree selection, and male-sterile parents (from Chaudhary et al. 1981).

Gm-1 gene conditions resistance to gall midge biotypes in Thailand and Andhra Pradesh, India, and the Gm-2 gene controls resistance to biotypes in Indonesia, Sri Lanka, and Orissa and Andhra Pradesh, India.

The inheritance of white-backed plant hopper resistance in rice is also controlled by the effects of both dominant and recessive genes. Wbph 1, in the cultivar 'N22,' and Wbph 2, in the cultivar 'ARC10239,' are inherited as dominant traits (Angeles et al. 1981, Sidhu et al. 1979). Wbph 3, in the cultivar 'ADR52,' and wbph 4, in the cultivar 'Podiwi-A8,' are inherited as dominant and recessive traits, respectively (Hernandez & Khush 1981). Whph 5 is a dominant trait for resistance in the cultivar 'NDaing marie' (IRRI 1984). Two cultivars from Pakistan have resistance to the white-backed plant hopper governed by Whph 1 and a recessive gene that segregate independently of one another (Nair et al. 1982). Gupta and Shukla (1986) determined that a recessive gene and an unidentified gene control resistance in the breeding line 274-A/TN1. Recessive gene effects for white-backed plant hopper resistance were also noted by Rapusas and Heinrichs (1985).

Resistance in rice to the yellow stem borer is polygenic. Resistance occurs in several different breeding lines developed at the International Rice Research Institute from crosses involving traditional cultivars with moderate resistance and breeding lines with higher levels of resistance (Khush 1984). A modified breeding strategy (Fig. 8.5) using a male-sterile female parent, recurrent selection, and pedigree selection is currently used to develop composite cultivars with increased levels of yellow stem borer resistance (Chaudhary et al. 1981).

8.3.6 Sorghum

Resistance in sorghum to the greenbug is controlled by a single gene that is expressed as incomplete dominance (Weibel et al. 1972). Conversely, resistance to the sorghum midge, *Contarinia sorghicola* (Coquillet), is inherited as a recessive trait and is controlled at two or more loci (Boozaya-Angoon et al. 1984, Rossetto & Igue 1983). Resistance is inherited from both additive and nonadditive genetic effects (Widstrom et al. 1984, Agarwal & Abraham 1985).

Research has also been conducted on resistance in sorghum to the sorghum shoot fly, *Atherigona soccata* Rond. Rana et al. (1981) evaluated several sources of shoot fly-resistant sorghum, and determined resistance to be controlled by additive polygenic effects. Resistance is expressed as a partially dominant trait at low to moderate shoot fly infestations. The additive component also increases in response to high shoot fly population, but the dominance component is unaffected (Borikar & Chopde 1980). Similar results were reported by Pathak and Olela (1983) for resistance in sorghum to the borer *Chilo partellus* (Swinhoe). Leaf trichomes also play a role in the shoot fly resistance of some sorghum cultivars (see Chapter 2). Gibson and Maiti (1983) evaluated shoot fly resistance in the progeny of crosses between pubescent and glaborous sorghum cultivars, and observed resistance to be expressed as a recessive trait conditioned by a single gene.

8.3.7 Wheat

Several interspecific crosses combined with backcrossing have been employed to transfer insect resistance into common wheat, *Triticum aestivum*. Knowledge of the genetics of resistance has been facilitated by the use of *monosomic, disomic* (normal chromosome complement), and *trisomic* (extra chromosome complement) plants with one or two chromosomes less than normal.

Patterson and Gallun (1977) determined that 'Knox 62' and 'Purdue 4835A4-6' wheats possess the H_6 gene for Hessian fly, *Mayetiola destructor* Say, biotype B resistance derived from PI94587. Resistance is dominant and linked to the H_3 gene that possesses resistance to Hessian fly biotype C. In related research, Gallun and Patterson (1977) crossed monosomic, Hessian fly-susceptible spring wheat cultivars to 'Purdue 4835A4-6,' a disomic, to determine the chromosomes responsible for fly resistance. Chromosome 5A carries resistance to Hessian fly biotypes A and B. Hessian fly resistance from *Triticum turgidum* is inherited independently of the H_5H_5 gene pair in the wheat cultivar 'Abe' (Stebbins et al. 1980). The gene for resistance is denoted as H_9. The H_3, H_6, and H_9 genes all occur on chromosome 5A, and both the H_6 and H_9 genes are linked to the H_3 gene.

Resistance in *T. turgidum* ('Ella' wheat) to Hessian fly biotype D was transferred into common wheat *Triticum aestivum*, by backcrossing, and is governed by two independent dominant genes (H_9 and H_{10}) (Stebbins et al. 1982, 1983). H_{11} is a resistance gene identified in the resistant progeny of crosses between the cultivar 'Riley' with *T. turgidium* (Stebbins et al. 1983). Hatchett et al. (1981) determined that resistance to Hessian fly biotype D in synthetic hexaploid wheats derived from tetraploid *Triticum* species and a diverse group of *T. tauschii* wheats was inherited as a dominant trait, involving the effect of a single gene. Resistance to Hessian fly biotype D in the cultivar 'Marquillo' is controlled by one partially dominant gene (Maas et al. 1987). 'Marquillo' resistance is inherited independently from the H_{12} gene for biotype D resistance in 'Lusco' wheat. The 'Marquillo' gene is also distinct from all other known genes for Hessian fly biotype D resistance.

Resistance to the greenbug in wheat is conditioned by a single recessive gene pair (Curtis et al. 1960). Greenbug resistance in *T. tauschii* (Joppa et al. 1980) is controlled by a single dominant gene. Hollenhorst and Joppa (1983) used the 'Chinese Spring' wheat monosomics to determine the chromosome location of the resistance genes in the greenbug resistant wheats 'Amigo' (*Triticum aestivum*) and 'Largo' (*T. turgidum* × *T. tauschii*). The resistance gene for 'Amigo' is located on chromosome 1A, and the gene for resistance in 'Largo' is located on chromosome 7D. Tyler et al. (1987) used the responses of the five greenbug biotypes to test plants and allelism tests to distinguish between the different types of greenbug resistance in wheat. Five different genes, designated as Gb1, Gb2, Gb3, Gb4, and Gb5, were determined to exist in the cultivars 'DS28A,' 'Amigo,' 'Largo,' 'CI17959' (*T. durum* × *T. taushii*), and 'CI17882' (CI15092 × *T. speltoides*), respectively.

8.4 FACTORS INFLUENCING INHERITANCE STUDIES

Plant phenological and genetic factors, environmental factors, and human physical resource limitations all influence the outcome of inheritance studies of plant resistance to insects. Genetic self-incompatibility and inbreeding depression in some cross-pollinated crops may also rapidly reduce the vigor of plant material rapidly in a breeding program and make this material difficult to transfer to donor parent plants.

Additive, recessive, dominant, or epistatic gene action and the number of gene loci for a given resistance factor also govern the rate of progress in breeding and inheritance research. Because the resistance of a cultivar may change with maturity, the progeny of crosses involving insect-resistant and insect-susceptible cultivars of different maturities may be hard to select. The amount of seed needed to determine adequately the inheritance of and the number of genes involved in resistance is usually large, and the process consumes large quantities of time and space. In order to control this problem, it is necessary to grow sizable plant populations to analyze inheritance, have controlled uniform infestations of insects, and design experiments that allow for the separate study of resistance mechanisms that require destructive sampling.

8.5 STRATEGIES FOR THE DEPLOYMENT OF GENES FOR RESISTANCE TO INSECTS

Strategies relying on the effects of either major or minor genes can be utilized to deploy insect-resistant cultivars into an insect pest management system. *Horizontal* or *polygenic resistance* (also called *field resistance*) utilizes a number of sources, each with a minor resistance gene, that are mixed and allowed to interbreed. Horizontal resistance is generally considered more stable than vertical resistance and is not readily overcome by resistance-breaking insect biotypes. Horizontal resistance is usually obtained by selection, random mating to obtain new gene combinations, and finally recurrent selection, to increase the frequency of resistance genes. Potential problems involved with horizontal resistance involve the lack of outcrossing in self-pollinated species of plants and the need for heavy uniform insect infestations. The use of male-sterile lines in breeding sorghum and rice cultivars with insect resistance has solved part of this problem.

Vertical or *monogenic resistance* relies on the effects of a single major gene that causes a high level of resistance to certain segments of the pest insect population. This resistance is less stable than horizontal resistance because it can be overcome by biotypes. There are several options for the use of vertical resistance. One option is to release one major gene, use it until it becomes ineffective, and make successive (sequential) releases of additional genes. *Sequential cultivar release* has been used for the deployment of genes in rice with brown planthopper resistance (Khush 1979). A second option is to *pyramid* two or more major genes in one cultivar. Gene pyramiding has been used successfully to protect wheat plantings against stem rust in Australia and in rice cultivars (IR36) with brown planthopper resistance.

Heinrichs (1986) and Khush (1984) describe the different gene deployment schemes currently used in rice cultivars with brown plant hopper resistance. Sequential release of cultivars controlled by monogenic resistance sources has been the strategy of necessity in use for several years. The cultivars currently in use in various areas of world rice production have resistance based on the six brown plant hopper resistance genes indicated previously. Now that a group of genes exists, some efforts to pyramid the effects of one or more genes are being pursued, but the effects of such combinations are untested. In Indonesia, *gene rotation* is being used to curb the development of brown plant hopper biotypes (Oka 1983). Cultivars with one set of genes are grown in the wet season production period and cultivars with a different set of genes grown in the dry season production period.

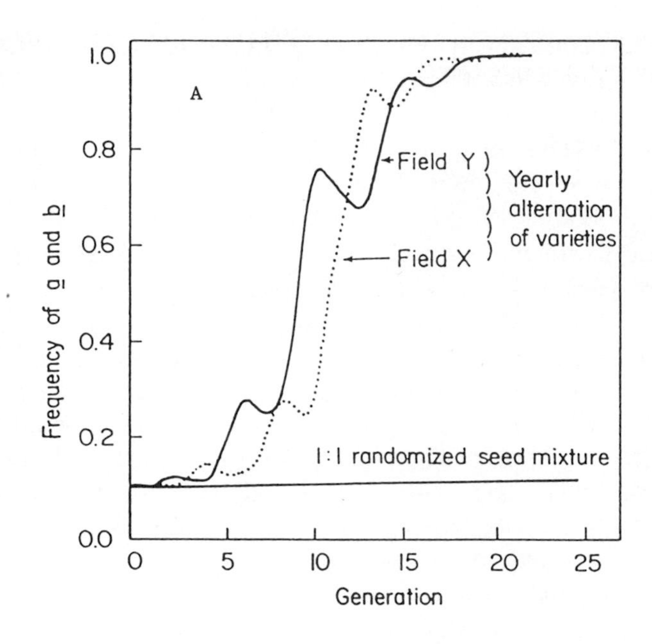

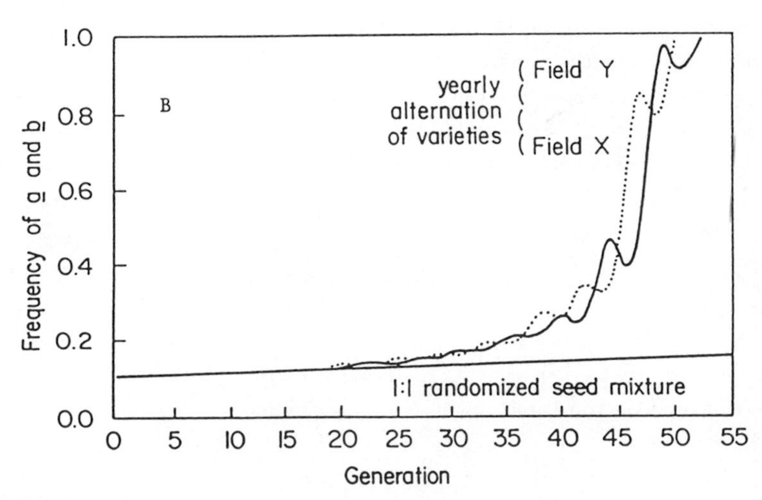

FIGURE 8.6 Effect of planting a two-gene pyramided Hessian fly, *Mayetiola destructor* (Say), resistant wheat cultivar in yearly alternation with a susceptible cultivar in comparison to a 1:1 mixture of resistant and susceptible plants grown annually. (*A*) Initial fly *a* and *b* allele frequencies 0.10. (*B*) Initial allele frequencies 0.04. (From Gould 1986a. Adapted with permission from Environ. Entomol. 15: 11–23. Copyright 1986, Entomological Society of America.)

A third option is the development and deployment of *crop multilines* (cultivars composed of different combinations of major and minor resistance genes) for protection against insects. Currently, geographic wheat multilines with resistance to the Hessian fly are grown in different wheat producing areas of the United States. Khush (1980) described a procedure to develop rice multilines with resistance to bacterial blight, grassy stunt disease (vectored by brown plant hopper), the brown plant hopper, and the green leafhopper.

Gould (1986a) used simulation models to determine how long an insect pest would require to adapt to two antibiotic plant resistance factors when they are deployed sequentially, as a cultivar multiline, or pyramided in a single cultivar. Because of the variations inherent to different insect pests and cropping systems, no one strategy is equally durable (long-lasting) over a wide variety of crop management systems. Model results indicate that sequentially released and two-gene multilines have lower resistance than pyramided cultivars, but their resistance is more durable. The durability of a pyramided cultivar is further enhanced by planting a percentage of the total crop in an insect–susceptible cultivar. More specific simulation model data, based on the interaction of the

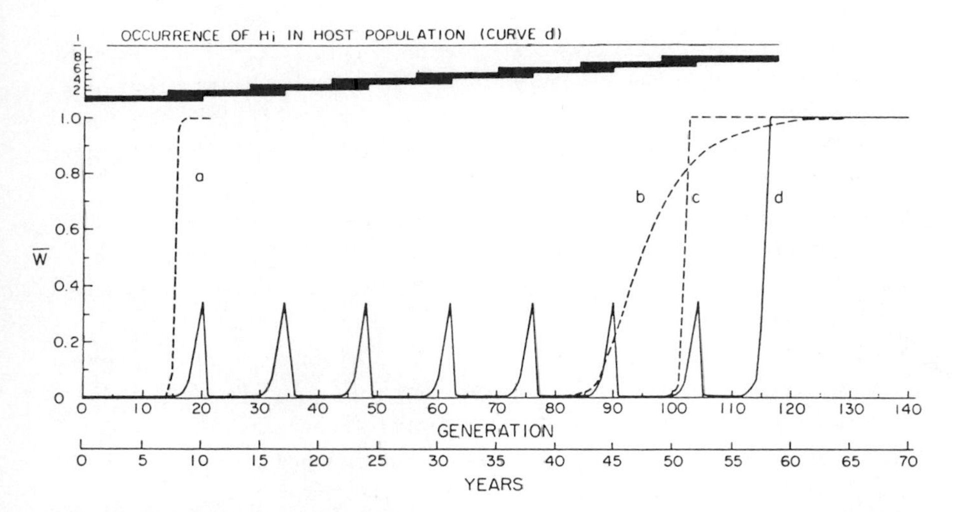

FIGURE 8.7 Effect of wheat genes with resistance to Hessian fly, *Mayetiola destructor* (Say) on relative mean fly fitness (W); (*a*) genes released simultaneously in one cultivar, (*b*) genes released simultaneously in eight different cultivars, (*c*) genes released sequentially in separate cultivars for 9 years each, (*d*) genes released sequentially in separate cultivars for 10 years each. H_i = each of the eight different resistance alleles in (*d*). From Cox and Hatchett 1986. Adapted with permission from Environ. Entomol. 15: 24–31. Copyright 1986, Entomological Society of America.)

Hessian fly with resistance in winter wheat (Gould 1986b), confirms the general predictions about the use of a combination of cultivars with two pyramided resistance genes and some totally susceptible plants (Fig. 8.6). Depending on the initial allele frequencies of Hessian fly populations, resistance is predicted to last from 90 to 400 fly generations (45–200 years).

In contrast, simulation modeling of all eight wheat genes involved in Hessian fly resistance generates somewhat different results concerning the lack of durability of pyramided gene cultivars, compared to the sequential release of cultivar release (Fig. 8.7) (Cox & Hatchett 1986). In these simulations, the sequential release of one gene every 4 years maintains resistance approximately 10 times longer than if the resistance of all eight genes were pyramided into a single cultivar and released simultaneously. Additional disadvantages of the release of a cultivar with several pyramided genes are the long-term effort required for cultivar development and the risk involved in the placement of all known resistance genes into a single cultivar.

REFERENCES

Agarwal, B. L. and C. V. Abraham. 1985. Breeding sorghum for resistance to shootfly and midge. In. *Proceedings of the International Sorghum Entomology Workshop*, July 15–21, 1984, Texas A & M University, College Station, TX. pp. 371–384.

Alston, F. H. and J. B. Briggs. 1968. Resistance to *Sappaphis devecta* (Wld.) in apple. Euphytica 17: 468–472.

Alston, F. H. and J. B. Briggs. 1970. Inheritance of hypersensitivity to rosy apple aphid, *Dysaphis plantaginea* in apple. Can. J. Genet. Cytol. 12: 257–258.

Angeles, E. R., G. S. Khush, and E. A. Heinrichs. 1981. New genes for resistance to whitebacked planthopper in rice. Crop Sci. 21: 47–50.

Angeles, E. R., G. S. Khush, and E. A. Heinrichs. 1986. Inheritance of resistance to planthoppers and leafhoppers in rice. In *Rice Genetics*. Proc. Intl. Rice Genetics Symp. May 27–31, 1985. Island Publ. Co., Manila, pp. 537–549.

Athwal, D. S. and M. D. Pathak. 1971. Genetics of resistance to rice insects. In *Rice Breeding*. IRRI, Los Banos, Philippines, pp. 375–386.

Auld, D. L., L. E. O'Keeffe, G. A. Murray, and J. L. Smith. 1980. Diallel analyses of resistance to the adult pea weevil in peas. Crop Sci. 20: 760–766.

Barnes, W. C. and F. P. Cuthbert, Jr. 1975. Breeding turnips for resistance to the turnip aphid. Hortic. Sci. 10: 59–60.

Bata, H. D., B. B. Singh, S. R. Singh, and T. A. O. Ladeinde. 1987. Inheritance of resistance to aphid in cowpea. Crop Sci. 27: 892–894.

Bird, L. S. 1982. Multi-adversity (diseases, insects and stresses) resistance (MAR) in cotton. Plant Dis. 66: 173–176.

Boozaya-Angoon, D., K. J. Starks, D. E. Weibel, and G. L. Teetes. 1984. Inheritance of resistance in sorghum, *Sorghum bicolor*, to the sorghum midge, *Contarinia sorghicola* (Diptera: Cecidomyiidae). Environ. Entomol. 13: 1531–1534.

Borikar, S. T. and P. R. Chopde. 1980. Inheritance of shoot fly resistance under three levels of infestation in sorghum. Maydica XXV: 175–183.

Burton, G. W., W. W. Hanna, J. C. Johnson, Jr., D. B. Leuck, W. G. Monson, J. B. Powell, H. D. Wells, and N. W. Widstrom. 1977. Pleiotropic effects of the *tr* trichomeless gene in pearl millet transpiration, forage quality and pest resistance. Crop Sci. 17: 613–616.

Chaudhary, R. C., E. A. Heinrichs, and G. S. Khush. 1981. Increasing the level of stem borer resistance through male sterile facilitated recurrent selection in rice. IRRN 8(5): 7–8.

Chaudhary, B. P., P. S. Srivastava, M. N. Shrivastava, and G. S. Khush. 1986. Inheritance of resistance to gall midge in some cultivars of rice. In *Rice Genetics*. Proc. Intl. Rice Genetics Symp. May 27–31, 1985. Island Publ. Co., Manila, pp. 523–527.

Chelliah, S. 1986. Genetics of resistance in rice to planthoppers and leafhoppers. In *Rice Genetics*. Proc. Intl. Rice Genetics Symp. May 27–31, 1985. Island Publ. Co., Manila, pp. 513–522.

Chiang, M. S. and M. Hudson. 1973. Inheritance of resistance to the European corn borer in grain corn. Can. J. Plant Sci. 53: 779–782.

Cox, T. S. and J. H. Hatchett. 1986. Genetic model for wheat/Hessian fly (Diptera: Cecidomyiidae) interaction: Strategies for deployment of resistance genes in wheat cultivars. Environ. Entomol. 15: 24–31.

Curtis, B. C., A. M. Schleuber, and E. A. Wood. 1960. Genetics of greenbug resistance in two strains of common wheat. Agron. J. 52: 599–602.

Cuthbert, F. P., Jr. and A. Jones. 1972. Resistance in sweet potatoes to Coleoptera increased by recurrent selection. J. Econ. Entomol. 65: 1655–1658.

DaCosta, C. P. and C. M. Jones. 1971. Cucumber beetle resistance and mite susceptibility controlled by the bitter gene in *Cucumis sativus* L. Science. 172: 1145–1146.

Dahms, R. G. 1972. Techniques in the evaluation and development of host-plant resistance. J. Environ. Qual. 1: 254–258.

Daubeny, H. A. 1966. Inheritance of immunity in the red raspberry to the North American strain of the aphid, *Amphorophora rubi* (Kitb.). Proc. Am. Soc. Horic. Sci. 88: 346–351.

dePonti, O. M. B. 1979. Resistance in *Cucumis sativus* L. to *Tetranychus urticae* Koch. 5. Raising the resistance level by the exploitation of transgression. Euphytica 28: 569–577.

Eenink, A. H., R. Groenwald, and F. L. Dieleman, 1982a. Resistance of lettuce (*Lactuca*) to the leaf aphid *Nasonovia ribis nigri* 1. Transfer of resistance from *L. virosa* to *L. sativa* by interspecific crosses and selection of resistant breeding lines. Euphytica 31: 291–300.

Eenink, A. H., F. L. Dieleman, and R. Groenwold. 1982b. Resistance of lettuce (*Lactuca*) to the leaf aphid *Nasonoria ribis nigri*. 2. Inheritance of the resistance. Euphytica 31: 301–304.

Elden, T. C. and J. H. Elgin, Jr. 1987. Recurrent seedling and individual-plant selection for potato leafhopper (Homoptera: Cicadellidae) resistance in alfalfa. J. Econ. Entomol. 80: 690–695.

Elden, T. C., J. H. Elgin, Jr., and J. F. Soper. 1986. Inheritance of pubescence in selected clones from two alfalfa populations and relationship to potato leafhopper resistance. Crop Sci. 26: 1143–1146.

Fatunla, T. and K. Badaru. 1983. Inheritance of resistance to cowpea weevil (*Callosobruchus maculatus*, Fabr.). J. Agric. Sci. Camb. 101: 423–426.

Fery, R. L. and F. P. Cuthbert, Jr. 1975. Inheritance of pod resistance to cowpea curculio infestation in southern peas. J. Hered. 66: 43–44.

Gallun, R. L. and F. L. Patterson. 1977. Monosomic analysis of wheat for resistance to Hessian fly. J. Hered. 68: 223–226.

Gibson, R. W. 1979. The geographical distribution, inheritance and pest-resisting properties of sticky-tipped foliar hairs on potato species. Potato Res. 22: 223–237.

Gibson, P. T. and R. K. Maiti. 1983. Trichomes in segregating generations of sorghum matings. I. Inheritance of presence and density. Crop Sci. 23: 73–75.

Glover, E. and B. Melton. 1966. Inheritance patterns of spotted alfalfa aphid resistance in *Zea* plants. N.M. Agric. Exp. Sta. Res. Rep. 127: 4.

Glover, D. V. and E. H. Stanford. 1966. Tetrasomic inheritance of resistance in alfalfa to the pea aphid. Crop Sci. 6: 161–165.

Gorz, H. J., G. R. Manglitz, and F. A. Haskins. 1979. Selection for yellow clover aphid and pea aphid resistance in red clover. Crop Sci. 19: 257–260.

Gould, F. 1986a. Simulation models for predicting durability of insect-resistant germplasm: A deterministic diploid, two-locus model. Environ. Entomol. 15: 1–10.

Gould, F. 1986b. Simulation models for predicting durability of insect-resistant germ plasm: Hessian fly (Diptera: Cecidomyiidae)-resistant winter wheat. Environ. Entomol. 15: 11–23.

Gracen, V. E. 1989. Breeding for resistance to European corn borer. In *Towards Insect Resistant Maize for the Third World: Proceedings of the International Symposium on Methodologies for Developing Host Plant Resistance to Maize Insects.* CIMMYT, Mexico, pp. 201–204.

Graham, J. H., R. R. Hill, Jr., D. K. Barnes, and C. H. Hanson. 1965. Effect of three cycles of selection for resistance to common leaf spot in alfalfa. Crop Sci. 5: 171–173.

Gupta, A. K. and K. K. Shukla. 1986. Sources and inheritance of resistance to whitebacked planthopper, *Sogatella furcifera* in rice. In *Rice Genetics*. Proc. Intl. Rice Genetics Symp. May 27–31, 1985. Island Publ. Co., Manila, pp. 529–535.

Hanson, C. H., J. H. Busbice, R. R. Hill, Jr., O. J. Hunt, and A. J. Oakes. 1972. Directed mass selection for developing multiple pest resistance and conserving germplasm in alfalfa. J. Environ. Qual. 1: 106–111.

Harlan, S. C. 1916. The inheritance of immunity to leaf blister mite (*Eriophyes gossypii* Banks) in cotton. West Indian Bull. 17: 162–166.

Hatchett, J. H., T. J. Martin, and R. W. Livers. 1981. Expression and inheritance of resistance to Hessian fly in synthetic hexaploid wheats derived from *Triticum tauschii* (Coss) Schmal. Crop Sci. 21: 731–734.

Heinrichs, E. A. 1986. Perspectives and directions for the continued development of insect-resistant rice varieties. Agric. Ecosyst. Environ. 18: 9–36.

Hernandez, J. E. and G. S. Khush. 1981. Genetics of resistance to whitebacked planthopper in some rice (*Oryza sativa* L.) varieties. Oryza 18: 44–50.

Hollenhorst, M. M. and L. R. Joppa. 1983. Chromosomal location of genes for resistance to greenbug in 'Largo' and 'Amigo' wheats. Crop Sci. 23: 91–93.

Ikeda, R. and C. Kaneda. 1981. Genetic analysis of resistance to brown planthopper, *Nilaparvata lugens* Stal, in rice. Jap. J. Plant Breed 31: 279–285.

Ikeda, R. and C. Kaneda. 1986. Genetic analysis of resistance to brown planthopper in rice. In *Rice Genetics*. Proc. Intl. Rice Genetics Symp. May 27–31, 1985. Island Publ. Co., Manila, pp. 505–513.

International Rice Research Institute (IRRI). 1984. Research highlights for 1983. Los Banos, Philippines, 121 pp.

Jennings, C. W., W. A. Russell, and W. D. Guthrie. 1974. Genetics of resistance in maize to first- and second-brood European corn borer. Crop Sci. 14: 394–398.

Johnson, A. L. and B. H. Beard. 1977. Sunflower moth damage and inheritance of the phytomelanin layer in sunflower achenes. Crop Sci. 17: 369–372.

Jones, A. and F. P. Cuthbert, Jr. 1973. Associated effects of mass selection for soil-insect resistance in sweet potato. J. Am. Soc. Hortic. Sci. 98: 480–482.

Jones, A., P. D. Dukes, and F. P. Cuthbert, Jr. 1976. Mass selection in sweet potato: breeding for resistance to insects and diseases and for horticultural characteristics. J. Am. Soc. Hortic. Sci. 101: 701–704.

Joppa, L. R., R. G. Timian, and N. D. Williams. 1980. Inheritance of resistance to greenbug toxicity in an amphiploid of *Triticum turgidum*/*T. tauschi*. Crop Sci. 20: 343–344.

Khush, G. S. 1978. Biology, techniques and procedures employed at IRRI for developing rice germplasm with multiple resistance to diseases and insects. Jap. Trop. Agric. Res. Ctr. Res. Ser. #11. 8 pp.

Khush, G. S. 1979. Genetics and breeding for resistance to the brown planthopper. In

Proc. Symp. Brown Plant Hopper. Threat to Rice Production in Asia. International Rice Research Institute, Los Banos, Philippines, pp. 321–332.

Khush, G. S. 1980. Breeding rice for multiple disease and insect resistance. In *Rice Improvement in China and Other Asian Countries*. International Rice Research Institute, Los Banos, Philippines, pp. 219–238.

Khush, G. S. 1984. Breeding rice for resistance to insects. Prot. Ecol. 7: 147–165.

Kilen, T. C., J. H. Hatchett, and E. E. Hartwig. 1977. Evaluation of early generation soybeans for resistance to soybean looper. Crop Sci. 17: 397–398.

Klenke, J. R., W. A. Russell, and W. D. Guthrie. 1986. Distributions for European corn borer (Lepidoptera: Pyralidae) ratings of S_1 lines from 'BS9' corn. J. Econ. Entomol. 79: 1076–1081.

Kornegay, J. L. and S. R. Temple. 1986. Inheritance and combining ability of leafroller defense mechanisms in common bean. Crop Sci. 26: 1153–1158.

Lakshminarayana, A. and G. S. Khush. 1977. New genes for resistance to the brown planthopper in rice. Crop Sci. 17: 96–100.

Livers, R. W. and F. L. Harvey. 1969. Greenbug resistance in rye. J. Econ. Entomol. 62: 1368–1370.

Luedders, V. D. and W. A. Dickerson. 1977. Resistance of selected soybean genotypes and segregating populations to cabbage looper feeding. Crop Sci. 17: 395–396.

Lyman, J. M. and C. Cardona. 1982. Resistance in lima beans to a leafhopper, *Empoasca krameri*. J. Econ. Entomol. 75: 281–286.

Maas III, F. B., F. L. Patterson, J. E. Foster, and J. H. Hatchett. 1987. Expression and inheritance of resistance of 'Marquillo' wheat to Hessian fly biotype D. Crop Sci. 27: 49–52.

Manglitz, G. R. and H. J. Gorz. 1968. Inheritance of resistance in sweetclover to the sweetclover aphid. J. Econ. Entomol. 61: 90–94.

Mehlenbacher, S. A., R. L. Plaisted, and W. M. Tingey. 1983. Inheritance of glandular trichomes in crosses with *Solanum berthaultii*. Am. Potato. J. 60: 699–708.

Mehlenbacher, S. A., R. L. Plaisted, and W. M. Tingey. 1984. Heritability of trichome density and droplet size in interspecific potato hybrids and relationship to aphid resistance. Crop Sci. 24: 320–322.

Nair, R. V., T. M. Masajo, and G. S. Khush. 1982. Genetic analysis of resistance to whitebacked planthopper in twenty-one varieties of rice, *Orzya sativa* L. Theor. Appl. Genet. 61: 19–22.

Nouri-Ghanbalani, G., D. L. Auld, L. E. O'Keeffe, and A. R. Campbell. 1978. Inheritance of resistance to the adult pea leaf weevil in Austrian winter peas. Crop Sci. 18: 858–860.

Oka, I. N. 1983. The potential for the integration of plant resistance, agronomic, biological, physical/mechanical techniques, and pesticides for pest control in farming systems. In L. W. Shemitt (ed.). *Chemistry and World Supplies: The New Frontiers, CHEMRAWN II.* Pergamon, Oxford, pp. 173–184.

Onukogu, F. A., W. D. Guthrie, W. A. Russell, G. L. Reed, and J. C. Robbins. 1978. Location of genes that condition resistance in maize to sheathcollar feeding by second-generation European corn borers. J. Econ. Entomol. 71: 1–4.

Panda, N. and E. A. Heinrichs. 1983. Levels of tolerance and antibiosis in rice varieties having moderate resistance to the brown planthopper, *Nilaparavata lugens* (Stal) (Homoptera: Delphacidae). Environ. Entomol. 12: 1204–1214.

Pathak, R. S. and J. C. Olela. 1983. Genetics of host plant resistance in food crops with special reference to sorghum stem-borers. Insect Sci. Appl. 4: 127–134.

Patterson, F. L. and R. L. Gallun. 1977. Linkage in wheat of the H_3 and H_6 genetic factors for resistance to Hessian fly. J. Hered. 68: 293–296.

Rana, B. S., M. G. Jotwani, and N. G. P. Rao, 1981. Inheritance of host plant resistance to the sorghum shootfly. Insect. Sci. Appl. 2: 105–110.

Rapusas, H. R. and E. A. Heinrichs. 1985. Whitebacked planthopper growth and development on rices with monogenic or digenic resistance. IRRN 10(5): 9–10.

Redden, R. J., P. Dobie, and A. Gatehouse. 1983. The inheritance of seed resistance to *Callosobruchus maculatus* F. in cowpea (*Vigna unguiculata* (L.) Walp.) I. Analyses of parental, F1, F2 and backcross seed generation. Aust. J. Agric. Res. 34: 681–695.

Redden, R. J., S. R. Singh, and M. J. Luckfahr. 1984. Breeding for cowpea resistance to bruchids at IITA. Prot. Ecol. 7: 291–303.

Rezual Kamin, A. N. M. and M. D. Pathak. 1982. New genes for resistance to green leafhopper, *Nephotettix virescens* (Distant) in rice, *Oryza sativa* L. Crop Prot. 1: 483–490.

Rossetto, C. J. and T. Igue. 1983. Heranca de resistencia variedade de sorgo AF28 a *Contarinia sorghicola* Coquillett. Bragantia 42: 211–219.

Russell, W. A., G. D. Lawrence, and W. D. Guthrie. 1979. Effects of recurrent selection for European corn borer resistance on other agronomic characters in synthetic cultivars of maize. Maydica 24: 33–47.

Sams, D. W., F. I. Lauer, and E. B. Radcliffe. 1976. Breeding behavior of resistance to green peach aphid in tuber-bearing *Solanum* germplasm. Am. Potato J. 53: 23–29.

Sanford, L. L. and T. L. Ladd, Jr. 1983. Selection for resistance to potato leafhopper in potatoes. III. Comparison of two selection procedures. Am. Potato J. 60: 653–659.

Sanford, L. L. and T. L. Ladd, Jr. 1987. Genetic transmission of potato leafhopper resistance from recurrent selection populations in potato *Solanum tuberosum* L. Gp. tuberosum. Am. Potato J. 64: 655–662.

Sastry, M. V. S. and P. S. Prakasa Rao. 1973. Inheritance of resistance to rice gall midge *Pachydiplosis oryzae* Wood-Mason. Curr. Sci. 42: 652–653.

Satyanarayanaiah, K. and M. V. Reddi. 1972. Inheritance of resistance to insect gall-midge (*Pachydiplosis oryzae*, Wood Mason) in rice. Andhra Agric. J. 19: 1–8.

Scott, D. R. 1977. Selection for *Lygus* bug resistance in carrot. Hort Sci. 12: 452.

Scott, G. E. and W. D. Guthrie. 1967. Reactions of permutations of maize double crosses to leaf feeding of European corn borers. Crop Sci. 7: 233–235.

Scott, G. E., F. F. Dicke, and G. R. Pesho. 1966. Location of genes conditioning resistance in corn to leaf feeding of the European corn borer. Crop Sci. 4: 444–446.

Sharma, G. C. and C. V. Hall. 1971. Cucurbitacin B and total sugar inheritance in *Cucurbita pepo* L. related to spotted cucumber beetle feeding. J. Am. Soc. Hortic. Sci. 96: 750–754.

Sidhu, G. S. and G. S. Khush. 1978. Genetic analysis of brown planthopper resistance in twenty varieties of rice, *Oryza sativa* L. Theor. Appl. Genet. 53: 199–203.

Sidhu, G. S., G. S. Khush, and F. G. Medrano. 1979. A dominant gene in rice for resistance to whitebacked planthopper and its relationship to other plant characteristics. Euphytica 28: 227–232.

Sifuentes, J. A. and R. H. Painter. 1964. Inheritance of resistance to western corn rootworm adults in field corn. J. Econ. Entomol. 57: 475–477.

Singh, D. P. 1986. *Breeding for Resistance to Diseases and Insect Pests.* Springer, New York, 222 pp.

Sisson, V. A., P. A. Miller, W. V. Campbell, and J. W. Van Duyn. 1976. Evidence of inheritance of resistance to the Mexican bean beetle in soybeans. Crop Sci. 16: 835–837.

Siwi, B. H. and G. S. Khush. 1977. New genes for resistance to the green leafhopper in rice. Crop Sci. 17: 17–20.

Smith, C. M. and C. A. Brim. 1979. Field and laboratory evaluations of soybean lines for resistance to corn earworm leaf feeding. J. Econ. Entomol. 72: 78–80.

Soper, J. F., M. S. McIntosh, and T. C. Elden. 1984. Diallel analysis of potato leafhopper resistance among selected alfalfa clones. Crop Sci. 24: 667–670.

Stebbins, N. B., F. L. Patterson, and R. L. Gallun. 1980. Interrelationships among wheat genes for resistance to Hessian fly. Crop Sci. 20: 177–180.

Stebbins, N. B., F. L. Patterson, and R. L. Gallun. 1982. Interrelationships among wheat genes H3, H6, H9, and H10 for Hessian fly resistance. Crop Sci. 22: 1029–1032.

Stebbins, N. B., F. L. Patterson, and R. L. Gallun. 1983. Inheritance of resistance of PI94587 wheat to biotypes B and D of Hessian fly. Crop Sci. 23: 251–253.

Takita, T. and H. Habibuddin. 1986. Variation and inheritance of resistance of rice to green leafhopper in Malaysia. Jap. Agric. Res. Quant. 19: 306–310.

Theurer, J. C., C. C. Blickenstaff, G. G. Mahrt, D. L. Doney. 1982. Breeding for resistance to the sugarbeet root maggot. Crop Sci. 22: 641–645.

Tyler, J. M., J. A. Webster, and O. G. Merkle. 1987. Designations for genes in wheat germplasm conferring greenbug resistance. Crop. Sci. 27: 526–527.

Velusamy, R., E. A. Heinrichs, and G. S. Khush. 1987. Genetics of field resistance of rice to brown planthopper. Crop Sci. 27: 199–200.

Weibel, D. E., K. J. Starks, E. A. Wood, Jr., and R. D. Morrison. 1972. Sorghum cultivars and progenies rated for resistance to greenbugs. Crop Sci. 12: 334–336.

Widstrom, N. W. 1972. Reciprocal differences and combining ability for corn earworm injury among maize single crosses. Crop Sci. 12: 245–247.

Widstrom, N. W. 1976. Graphic interpretation of regressions of F_1 on mid-parent and the application to anomalous data. J. Hered. 67: 54–58.

Widstrom, N. W. 1989. Breeding methodology to increase resistance in maize to corn earworm, fall armyworm, and maize weevil. In *Towards Insect Resistant Maize for the Third World: Proceedings of the International Symposium on Methodologies for Developing Host Plant Resistance to Maize Insects*. CIMMYT, Mexico, pp. 209–219.

Widstrom, N. W. and J. J. Hamm. 1969. Combining abilities and relative dominance among maize inbreds for resistance to earworm injury. Crop Sci. 9: 216–219.

Widstrom, N. W. and W. W. McMillian. 1973. Genetic factors conditioning resistance to earworm in maize. Crop Sci. 13: 459–461.

Widstrom, N. W., B. R. Wiseman, and W. W. McMillian. 1972. Resistance among some maize inbreds and single crosses to fall armyworm injury. Crop Sci. 12: 290–292.

Widstrom, N. W., B. R. Wiseman, and W. W. McMillian. 1982. Responses to index selection in maize for resistance to ear damage by the corn earworm. Crop Sci. 22: 843–846.

Widstrom, N. W., B. R. Wiseman, and W. W. McMillian. 1984. Patterns of resistance in sorghum to the sorghum midge. Crop Sci. 24: 791–793.

Wilson, F. D. and B. W. George. 1979. Combining ability in cotton for resistance to pink bollworm. Crop Sci. 19: 834–836.

Wilson, F. D. and B. W. George. 1983. A genetic and breeding study of pink bollworm resistance and agronomic properties in cotton. Crop Sci. 23: 1–4.

Wilson, F. D. and J. A. Lee. 1971. Genetic relationships between tobacco budworm feeding response and gland number in cotton seedlings. Crop Sci. 11: 419–421.

Wilson, F. D. and J. N. Smith. 1977. Variable expressivity and gene action of gland-determining alleles in *Gossypium hirsutum* L. Crop Sci. 17: 539–43.

Wright, R. J., M. B. Dimock, W. M. Tingey, and R. L. Plaisted. 1985. Colorado potato beetle (Coleoptera: Chrysomelidae): Expression of resistance in *Solanum berthaultii* and interspecific potato hybrids. J. Econ. Entomol. 78: 576–582.

9 Insect Biotypes That Overcome Plant Resistance

9.1 SIGNIFICANCE OF BIOTYPES FOR THE DEVELOPMENT OF INSECT-RESISTANT CULTIVARS

The protective properties of insect-resistant cultivars may be overcome by the development of resistance-breaking biotypes that possess an inherent genetic capability to overcome plant resistance. Typically, biotypes develop as a result of selection from the parent population in response to exposure to the resistant cultivar. The intensity and duration of this selection and the initial frequency of insect genes for overcoming resistance are the major factors that govern the rate of biotype development. It is also possible that the genetic capability of an insect to overcome plant resistance may be so great that the value of the resistance is nullified before the resistant cultivar is used on a widespread geographic basis. This phenomenon occurred in populations of the cabbage aphid *Brevicornye brassicae*, exposed to resistant cultivars of brussels sprouts (Dunn & Kempton 1972).

Biotypes have been defined as populations within an insect species that vary in their ability to utilize a crop plant (Gallun & Khush 1980). Some insect biotypes, such as those of the brown plant hopper, *Nilaparvata lugens* Stal, and the greenbug, *Schizaphis graminum* Rondani, are distinguishable by morphological differences (Saxena & Rueda 1982, Starks & Burton 1972). Others, such as those of the pea aphid, *Acyrthosiphon pisum* (Harris), and the English grain aphid, *Macrosiphum avenae* (F.), exhibit different color forms or morphs (Auclair 1978, Lowe 1981). Differences in insecticide susceptibility also exist between biotypes of the greenbug (Teetes et al. 1975), the spotted alfalfa aphid, *Therioaphis maculata* (Buckton) (Nielson & Don 1974), and the woolly apple aphid, *Eriosoma lanigerum* (Haussmann) (Sen Gupta & Miles 1975). Eastop (1973) cites these and other criteria, such as insect feeding behavior, host preference, and virus transmission ability, (Thottappilly et al. 1972) as a means of differentiating between insect biotypes.

9.2 TAXONOMIC DISTRIBUTION OF BIOTYPES

Based on the previously mentioned criteria, 14 insects exhibit biotypes with the ability to overcome genetic plant resistance to insects (Table 9.1). Nine of the 14 are aphid species, in which parthenogenetic reproduction contributes greatly to the successful development of resistance-breaking biotypes. Two of the other five species, the Hessian fly, *Mayetiola destructor* Say, and the rice gall midge, *Orseolia oryzae* (Wood-Mason), are sexually dimorphic Diptera with high reproductive potentials that infest large cereal monocultures. The brown plant hopper, green leafhopper, *Nephotettix virescens* (Distant), and rice green leafhopper, *Nephotettix cincticeps* Uhler, occur continuously on large rice monocultures in much of southern and Southeast Asia. For additional general information on aphid biotypes, see the bibliography of Webster and Inayatulluh (1985).

TABLE 9.1 Insects Developing Biotypes in Response to Plant Resistance

Crop	Insect	Number of Biotypes	References(s)
Alfalfa	*Acyrthosiphon pisum*, pea aphid	4	Auclair 1978, Frazer 1972
	Therioaphis maculata, spotted alfalfa aphid	6	Nielson and Lehman 1980
Apple	*Eriosoma lanigerum*, woolly apple aphid	2	Sen Gupta and Miles 1975
	Dysaphis devecta, rosy leaf curling aphid	3	Alston and Briggs 1977
Corn	*Rhopalosiphum maidis*, corn leaf aphid	5	Painter and Pathak 1962, Singh and Painter 1964, Wild and Feese 1973
Raspberry	*Amphorophora rubi*, raspberry aphid	4	Briggs 1965, Keep and Knight 1967
Rice	*Nephotettix cincticeps*, green rice leafhopper	2	Sato and Sogawa 1981
	Nephotettix virescens, green leafhopper	3	Heinrichs and Rapusas 1985, Takita and Hashim 1985
	Nilaparvata lugens, brown plant hopper	4	Heinrichs 1986, Kalode et al. 1975, Verma et al. 1979
	Orseolia oryzae, rice gall midge	4	Heinrichs and Pathak 1981
Vegetables	*Brevicoryne brassicae*, cabbage aphid	2–4	Dunn and Kempton 1972, Lammerink 1968
Wheat	*Macrosiphum avenae*, English grain aphid	3	Lowe 1981
	Mayetiola destructor, Hessian fly	8	Gallun and Reitz 1971, Sosa 1981
	Schizaphis graminum, greenbug	8	Harvey and Hackerott 1969, Kindler and Spomer 1986, Porter et al. 1982, Teetes et al. 1975, Wood 1961, Puterka et al. 1988

9.3 CASE HISTORIES OF THE DEVELOPMENT OF BIOTYPES

9.3.1 Fruit Crops

Two biotypes of the woolly aphid exist on the foliage of different apple clones in south Australia. The Clare (native) biotype is avirulent to resistant apple cultivars such as 'Northern Spy,' but the Blackwood biotype is virulent to several cultivars normally resistant to the Clare biotype (Sen Gupta & Miles 1975). Resistance in English apple cultivars to the rosy leaf curling aphid, *Dysaphis devecta* (Walker), follows a gene for gene relationship (Alston & Briggs 1977). Four resistance genes govern resistance to three aphid biotypes. Biotypes 1 and 2 are avirulent (nondamaging) on 'Cox's Orange Pippin,' which is protected by the gene denoted as Sd_1. Resistance to biotype 1 exists in 'Northern Spy,' only via the Sd_2 gene. A single gene (Sd_3) in *Malus robusta* and *Malus zumi* selections controls resistance to biotype 3. Several cultivars also possess precursor genes without which the biotype-specific genes for resistance are ineffective.

Four biotypes of the rubus aphid, *Amphorophora rubi* (Kalt.), exist on different cultivars of raspberry in England (Briggs 1965), and genes for resistance to biotypes 1, 2, and 3 have been identified (Table 9.2).

TABLE 9.2 Genes from *Rubus* Germplasm for Resistance to the Rubus Aphid, *Amphorophora rubi* (Kalt.)[a,b]

	Biotype		
Gene(s)	1	2	3
A_1	R	S	R
A_2	S	R	S
A_3A_4	S	R	S
A_5	R	S	S
A_6	R	S	S
A_7	R	S	S
$A_1 + A_2$	R	R	R
$A_1 + A_3$	R	R	R
$A_1 + A_4$	R	S	R

[a]From Keep and Knight (1967) Reprinted with permission from Euphytica 16: 209–214. Copyright 1967, Nijhoff Publishing Co.
[b]R = resistant; S = susceptible.

9.3.2 Legumes

As early as 1942, Harrington (1943) recognized differences in populations of the pea aphid. Cartier (1959, 1963) demonstrated the existence of three pea aphid biotypes on peas in the field and greenhouse. Frazer (1972) recognized five pea aphid biotypes, based on color, weight, and differential survival on several legume host plants. Auclair (1978) determined that morphological differences exist between two of the three pea aphid biotypes in North America, and that a fourth biotype exists in a pink color morph from France. Differences among the four biotypes are also evident when they are reared on an artificial diet.

The first biotype of the spotted alfalfa aphid was identified by Pesho et al. (1960) in Arizona. Since then, biotypes A, B, E, F, and H in Arizona and a Nebraska (N) biotype have been elucidated, based on differential responses to clones of 'Moapa' alfalfa (Table 9.3) (Nielson & Lehman 1980). Biotypes C, D, and G are resistant to organophosphorus insecticides (Nielson & Don 1974). Biotype development in the spotted alfalfa aphid also appears to have followed a gene for gene progression, although genetic analysis of alfalfa is difficult because of the heterozygosity and tetraploid conditions.

TABLE 9.3 Resistance of 'Moapa' Alfalfa Clones to Biotypes of the Spotted Alfalfa Aphid, *Therioaphis maculata* (Buckton)[a,b]

Alfalfa Clone	Aphid Biotype[c]					
	A	B	E	F	H	N[d]
C-903	S	R	S	S	S	R
C-904	R	R	S	I	I	R
C-905	R	R	S	S	S	R
C-906	R	R	R	S	S	R
C-907	R	R	R	R	S	R
C-908	S	R	S	S	S	R
C-909	R	R	R	I	I	R
C-910	R	R	R	I	S	R
C-911	S	R	S	S	S	S

[a]Adapted from Nielson and Lehman (1980) with permission of John Wily & Sons, Inc.
[b]R = resistant; I = Intermediate; S = susceptible.
[c]Biotypes C, D, and G are resistant to organophosphorus insecticides.
[d]Nebraska population.

9.3.3 Maize

Biotype development in the corn leaf aphid, *Rhopalosiphum maidis* (Fitch), was first detected by Cartier and Painter (1956), who observed an aphid population (KS-2) capable of surviving on sorghum cultivars previously resistant to the parent aphid population (KS-1). KS-3, a biotype capable of infesting wheat, and KS-4, a biotype capable of reproduction on barley at low (13°C) temperature, were detected in further research (Painter & Pathak 1962, Singh & Painter 1964). KS-5, detected in aphid field populations by Whalon et al. (1973), was so named because of its high temperature (29°C) tolerance and ability to feed on the wheat species *Triticum timopheevi*, which is resistant to the other four biotypes.

9.3.4 Rice

As indicated previously, biotype development has occurred in several rice insect pests owing to the continuous production of rice in large-scale monocultures. Three Southeast Asian biotypes and at least two southern Asian biotypes of the brown plant hopper exist because of their abilities to destroy different rice cultivars with specific genes for hopper resistance (Table 9.4).

The development and release of brown plant hopper-resistant cultivars by the International Rice Research Institute has followed a pattern of sequential release, using sources of monogenic resistance. 'IR 26,' the first cultivar released, is susceptible to the brown plant hopper in India (Verma et al. 1979). This was the first indication of the existence of brown plant hopper biotypes.

TABLE 9.4 Greenhouse Reaction of Rice Cultivars to Damage by Biotypes of the Brown Plant Hopper, *Nilaparvata lugens* Stal[a,b]

Cultivar	Gene(s) for Resistance	Philippines Biotype 1	Philippines Biotype 2	Philippines Biotype 3	South Asia Biotype Bangladesh	South Asia Biotype India	South Asia Biotype Sri Lanka
TN 1	None	S	S	S	S	S	S
IR 26	Bph 1	R	S	R	S	S	S
ASD 7	bph 2	R	R	S	S	S	S
Rathu Heenati	Bph 3	R	R	R	R	S	R
Babewee	bph 4	R	R	R	R	R	R
ARC 10550	bph 5	S	S	S	R	R	R
Ptb 33	2?	R	R	R	R	R	—
Sinna Sivappu	2?	R	R	R	R	R	R

[a]From Heinrichs 1986. Adapted with permission from Agric. Ecosyst. Environ. 18:9–36. Copyright 1986, Elsevier Science Publishers.
[b]R = resistant; S = susceptible.

Selection studies by Pathak and Heinrichs (1982) isolated a Philippine population unable to destroy the cultivars 'IR8,' 'IR20,' 'IR22,' and 'IR24' (biotype 1); a population damaging to 'IR26,' 'IR28,' and 'IR30' (biotype 2); and a population that destroys 'IR32,' 'IR36,' and 'IR38' (biotype 3). The dominant gene Bph 1 in 'IR26' confers resistance on Philippine biotypes 1 and 3 (Table 9.4). 'ASD7' contains the recessive gene bph 2, and is resistant to biotypes 1 and 2. Bph 3 in the cultivar 'Rathu Heenati' and bph 4 in the cultivar 'Babawee' provide resistance to all three Southeast Asian biotypes. Two genes for resistance to all three biotypes also exist in the cultivars 'Ptb 33' and 'Sinna Sivappu' (Table 9.4). After cultivation of IR 26 in the Philippines and Indonesia for 2 years, biotype 2 became predominant. In the meantime, 'IR 36' was developed, containing the bph 2 gene and biotype 2 resistance. Since its release in 1975, 'IR 36' has been cultivated on about 10 million hectares in the Philippines, Indonesia, the Solomon Islands, and Vietnam, without confirmed reports of biotype development. The stability of resistance in 'IR 36' is thought to be due to the effects of several minor genes, in addition to the bph 2 gene, to which biotype 2 has not adapted.

In addition to cultivar-specific or sympatric biotypes, allopatric brown plant hopper biotypes also occur in different geographic areas. Populations in Bangladesh, Sri Lanka, and southern India exhibit different responses from those of the Southeast Asian biotypes. Biotypes at all three locations are virulent to the Bph 1 and bph 2 genes (Table 9.4). In addition, the Hyderbad, India biotype is unaffected by the Bph 3 resistance gene in 'Rathu Heenati.' The digenic resistance in 'Ptb 33' and 'Sinna Sivappu' and the bph 4 gene in 'Babawee' also impart resistance to all southern Asian biotypes. ARC 10550, a cultivar susceptible to all Southeast Asian biotypes, is resistant to all southern Asian biotypes (Table 9.4). The results of Lin and Huang (1981) indicate that five brown plant hopper biotypes exist in the Republic of Taiwan. Entomologists throughout Asia are now attempting to differentiate the southern Asian and Southeast Asian biotypes through cooperative research.

Brown plant hopper biotypes show considerable variation between one another in both physiological and behavioral aspects of feeding. Paguia et al. (1980) demonstrated differential feeding activity of the three Philippine biotypes, based on the weight of honeydew collected from each of the three Philippine biotypes. Sogawa (1978a) determined distinct differences between the three Philippine biotypes in the amounts of several amino acids excreted in hopper honeydew. Brown plant hopper biotypes can also be distinguished by methods other than preferential feeding on different rice cultivars. A distinct electrophoretic variant is distinguishable for Philippine biotype 2 (Sogawa

1978b), based on esterase banding patterns. Morphological characteristics on the rostrum, legs, and antennae of both male and female hoppers also show distinct differences among the three Philippine biotypes (Saxena & Rueda 1982, Saxena & Barrion 1983).

Claridge and Den Hollander (1982) determined that each of the three Southeast Asian biotypes of the brown plant hopper can be effectively converted from themselves to each of the other two biotypes, depending on the rice cultivar on which they are reared. The results of Claridge and Den Hollander (1983), Den Hollander and Pathak (1981), and Guo-rui et al. (1983) suggest that brown plant hopper virulence on different rice cultivars is highly variable and may be controlled by a system of polygenes. For this reason, brown plant hopper biotypes are most accurately described by inclusion of cultivar and country of origin (Claridge & Den Hollander 1983).

Biotypes of the rice gall midge were first studied in India by Kalode et al. (1976). As the planting of gall midge resistant cultivars has increased, so apparently has the development of gall midge biotypes. Biotypes from Thailand, Indonesia, and India exist, based on their development and survival on a set of cultivar groups from India and Thailand (Table 9.5) (Heinrichs & Pathak 1981). All cultivar groups except those from the Eswarakora group are resistant to the Orissa, India biotype. Conversely, the Eswarakora and Ptb cultivar groups are resistant to the Andhra Pradesh, India biotype. The cultivar groups react in an opposite manner to the Indonesian biotype. The Eswarakora and Ptb groups are susceptible, but the *Oryza branchyantha*, Siam 29, and Leuang 152 groups are resistant. The Thailand biotype appears to

Table 9.5 Reaction of Rice Cultivars and Wild Rice, *Oryza brachyantha*, to Biotypes of the Gall Midge, *Orseolia oryzae* (Wood-Mason), from India, Indonesia, and Thailand[a]

	Cultivar Group and Country of Origin				
Biotype Country of Origin	Ptb (India)	*O. brachyantha* (India)	Siam 29 (Thailand)	Leuang 129 (Thailand)	Eswarakora (India)
Orissa, India	R	R	R	R	S
Andhra Pradesh, India	R	S	S	S	R
Indonesia	S	R	R	R	S
Thailand	R	S	S	S	S

[a]From Heinrichs and Pathak 1981. Reprinted permission from Insect Sci. Appl. 1:123–132. Copyright 1981, ICIPE Science Press.
[b]R = resistant; S = susceptible.

possess the highest degree of virulence, for it is capable of successfully utilizing all cultivar groups, except the Ptb group from India. Morphological variation exists between the gall midge biotypes from India and Thailand. Gall midges from Thailand are larger and have fewer large abdominal setae than those from India (Hidaka et al. 1977).

A gene for gene relationship also exists between biotypes of the green leafhopper and genes for hopper resistance in several rice cultivars. Takita and Hashim (1985) reared hoppers on five different resistant cultivars and were able to produce a resistance breaking biotype on each. Heinrichs and Rapusas (1985) obtained similar results and determined that minor genes are a factor in determining hopper cross resistance in different cultivars. As with the brown plant hopper, populations of the green leafhopper in different geographic locations have varying virulence patterns. Genes from cultivars that are resistant in the Philippines and Indonesia are susceptible to hopper populations in Bangladesh, India, and Malaysia (Heinrichs 1986). Biotypic variation has also been determined between populations of the green rice leafhopper in Japan by Saito and Sogawa (1981).

9.3.5 Vegetables

Biotypes of the cabbage aphid, occur on different cultivars of rape, *Brassica napus* in New Zealand (Lammerink 1968) and as mentioned in Section 9.1, on brussels sprouts clones in England (Dunn & Kempton 1972).

9.3.6 Wheat

More research has been conducted on biotype development in the greenbug and the Hessian fly than on any other insects forming biotypes in the United States. A large part of the development of greenbug biotypes is due to the ability of the greenbug to infest barley, oats, rye, sorghum, wheat, and numerous grasses (Michels 1986). Although there were no specific biotype designations, Dahms (1948) observed differences in Mississippi and Oklahoma greenbug populations as early as 1946. Wood (1961) named the first greenbug biotype B, owing to its ability to overcome the resistance of 'Dickinson 28A' wheat to biotype A (Table 9.6). Biotype B also has the ability to tolerate higher temperatures than biotype A (Singh & Wood 1963). In the late 1960s, greenbug populations became severely damaging in sorghum producing areas of the southwestern United States, and the sorghum biotype (C) developed (Harvey & Hackerott 1969) (Table 9.6). With increased insecticide usage on

TABLE 9.6 Characteristics and Dates Occurrence of Biotypes of the Greenburg, *Schizaphis graminum* (Rondani)[a]

Biotype	Occurrence	Identifying Characteristics	Reference (s)
A	Unknown	Virulent to wheat	Wood 1961
B	1961	Virulent to wheat	Wood 1961
C	1968	Expanded host range to and became a pest of sorghum	Harvey and Hackerott 1969
D	1974	Resistant to disulfoton insecticide	Peters et al. 1975, Teetes et al. 1975
E	1980	Virulent to biotype C-resistant 'Amigo' wheat, morphometrically and physiologically different from biotype C	Campbell and Dreyer 1985; Dreyer and Campbell 1984; Fargo 1986; Inayatullah et al. 1987a, b; Porter et al. 1982
F	1985	Able to feed on Canada bluegrass; lacks a dorsal stripe	Kindler and Spomer 1986
G	1988	Avirulent on 'Wintermalt' barley	Puterka et al. 1988
H	1988	Avirulent on sorghum; virulent on 'Post' 'barley	Puterka et al. 1988

[a]Adapted with permission from Michels 1986.

sorghum in Texas and Oklahoma, biotype D, resistant to the insectide disulfoton developed (Teetes et al. 1975, Peters et al. 1975).

In the late 1970s, biotype E, capable of feeding on the biotype C-resistant wheat 'Amigo,' developed in the high plains of Texas (Porter et al. 1982) (Table 9.6). Some biotype C-resistant wheat cultivars have retained antixenosis resistance to biotype E, but most barley and sorghum cultivars are susceptible (Starks et al. 1983). Biotype E resistance also exists in two sorghum plant introductions with the bloomless character (Starks et al. 1983) and in wheat germplasm with wheat streak mosaic virus resistance (Tyler et al. 1985). Biotype E greenbugs exhibit normal feeding behavior and fecundity on wheat cultivars resistant to biotype B (Niassy et al. 1987). Biotype E also possesses physiological superiority to biotype C, owing to increased activity of the digestive enzymes pectin methyl esterase and polysaccharase (Campbell &

Dreyer 1985, Dreyer & Campbell 1984) that enable biotype E to reproduce at approximately twice the rate of biotype C (Montllor et al. 1983).

There are significant morphometric differences among greenbug biotypes, the most prominent of which is the length of the first flagellar segment (Inayatullah et al. 1987a). Biotype E individuals also require a longer scotophase to induce the production of males than do biotype C individuals (Eisenbach & Mittler 1987). In contrast to other biotype-forming species of insects, no differences exist among greenbug biotypes B, C, or F for the electrophoretic banding patterns of several enzymes (Beregovoy & Starks 1986). Inayatullah et al. (1987b) used multivariate statistical analyses to study the morphometric characteristics of males and different female morphs of greenbug biotypes B, C, and E. Mahalanobis distance analysis of the data indicates that biotype E is closely related to biotype C, and that biotype C is closely related to biotype B.

A sixth greenbug biotype (F), isolated from turf grass in Ohio, differs from biotype E in its ability to kill Canada bluegrass and in its lack of a dorsal stripe (Kindler & Spomer 1986). Biotypes G and H were identified by Puterka et al. (1988). Both have the appearance of biotype F, but differ in their reactions with different small grain hosts. Biotype G can effectively utilize all known sources of greenbug resistance in wheat, but has no effect on 'Wintermalt' barley, which is normally susceptible to greenbugs. Biotype H has the same effect on wheat cultivars as biotype E, but can feed effectively on 'Post' barley, which is normally resistant to all greenbug biotypes. Populations of greenbugs from Oklahoma, Kentucky, and Maryland may also be biotypes (Kindler & Spomer 1986), but their genetic authenticity has not been documented. Different genotypes of greenbugs also occur on sorghum (pale green biotype) and oat (dark green biotype) in Argentina (Chidichimo et al. 1987).

The development of wheat cultivars in the United States began in New York in the early 1900s. However, the development of different biotypes of the Hessian fly has closely paralleled the development of wheat cultivars. The first recognition of Hessian fly biotypes was by Painter (1930), who referred to them as biological strains (biotypes have also been referred to as host-specific races). Ten different Hessian fly biotypes currently exist, and they are differentiated by how various combinations of four genes for virulence (the ability to overcome resistance) interact with four genes for resistance to Hessian flies in different wheat cultivars (Table 9.7). The interaction between Hessian fly biotypes and the genes for fly resistance in different wheat cultivars is the most thoroughly studied gene for gene relationship between an insect and its host plant. Virulence in a Hessian fly biotype depends on the existence of the homozygous

TABLE 9.7 Interaction of Genes for Resistance in Wheat Cultivars with Virulence Genes in Biotypes of the Hessian Fly, *Mayetiola destructor* (Say)[a]

	Biotype and Gene(s) for Virulence									
Wheat Cultivar and Gene(s) for Resistance	Great Plains (none)	A (s)	B (s, m)	C (s, k)	D (s, m, k)	E (m)	F (k)	G (m, k)	J (s, m, a)	L (s, m, k, a)
Turkey (none)	S	S	S	S	S	S	S	S	S	S
Seneca (H_7, H_8)	R	S	S	S	S	R	R	R	S	S
Monon (H_3)	R	R	S	R	S	S	R	S	S	S
Knox 62 (H_6)	R	R	R	S	S	R	S	S	R	S
Arthur 71 (H_5)	R	R	R	R	R	R	R	R	S	S

[a]Adapted from Gallun and Khush 1980, with permission of John Wiley & Sons, Inc.
[b]R = resistant; S = susceptible.

recessive condition in the virulence gene of the fly at a locus corresponding to a specific dominance gene for resistance (Gallun 1977). Thus in each instance of cultivar resistance (fly avirulence) in Table 9.7, the particular Hessian fly biotype in question lacks the homozygous recessive condition at the virulence gene locus.

The Great Plains biotype, isolated in western Kansas, carries the homozygous recessive condition for virulence only in association with the cultivar 'Turkey,' and lacks this condition when attempting to survive on cultivars with the H_3, H_5, H_6, or H_7 and H_8 genes. Biotype A, found in the eastern wheat growing areas of Indiana, Missouri, and Tennessee, is similar to the Great Plains biotype, but is homozygous recessive for virulence on the cultivar 'Seneca' (H_7 and H_8 genes). Biotype B is also found in several of the north central states and is the major biotype in Indiana. Biotype B possesses an additional virulence gene and is capable of breaking the resistance of the H_3 gene in the cultivar 'Monon.' Biotype C is also found in the north central region of the United States, but has the opposite reaction of biotype B to the H_3 and H_6 genes in 'Monon' and 'Knox 62' wheat, respectively. Biotype D, found only in Indiana in low density, is similar to biotype B, but is also virulent to the H_6 gene of 'Knox 62.' Biotype E was detected in Georgia by Hatchett (1969). Biotype E is similar to biotype C, but lacks the virulence condition when cultured on 'Seneca' (H_7 and H_8 genes).

Biotypes F and G were both created in the laboratory experimentally. Biotype F, the result of crossing of biotype C and the Great Plains biotype, has some traits of each of its parents. Biotype F reacts similarly to biotype C in response to the H_3, H_5, and H_6 genes, and similarly to the Great Plains biotype in response to the H_7 and H_8 genes. Biotype G, a cross between biotype D and the Great Plains biotype, is similar to the Great Plains biotype in its avirulent reaction to the H_7 and H_8 genes, and similar to biotype D in reaction to the H_3, H_5, and H_6 genes (Woottipreecha 1971). Biotypes J and L have developed in response to the resistance in the cultivar 'Arthur 71,' which possesses the H_5 gene from the cultivar 'Abe.' Both were detected in field populations in Indiana (Sosa 1981), although biotype L has also been selected in the greenhouse (Sosa 1978). The two biotypes are both virulent to the H_5 gene, but differ in response to the H_6 gene, which is resistant to biotype J (Table 9.7).

9.4 FACTORS CONTRIBUTING TO BIOTYPE SELECTION

From the preceding discussions, a general concept emerges about the types of pest management practices necessary to avoid the selection of insect biotypes. Although many studies have been conducted to determine the categories of insect resistance operating in food plants, few have contrasted and compared the advantages and disadvantages of different resistance categories during the exposure of a resistant cultivar to several generations of the pest insect. As discussed in Chapters 2, 3, and 4, the tolerance, antibiosis, and antixenosis categories of resistance differ in their effects on pest insect populations. Combinations of different resistance categories have effects that are more beneficial than the effect of individual categories of resistance. Simulation model studies by Kennedy et al. (1987) indicate that crop cultivars with combinations of antibiosis and antixenosis are less prone to biotype development by polyphagous insects than cultivars with only antibiosis. Thus combining different categories of resistance into one cultivar may also help slow the development of biotypes.

Information presented in Chapter 4 also demonstrates how cultivars resistant to insects by means of tolerance are normally more stable than those with antixenosis or antibiosis resistance. Several insect biotypes have developed in response to the effects of single genes whose effects are manifested as high levels of antixenosis and/or antibiosis. Therefore, one effective management practice is the use of a cultivar with a broad genetic base of resistance. This is accomplished by developing cultivars with horizontal resistance (Chapter 8). Horizontal resistance now exists in some alfalfa cultivars resistant to the western biotypes of the spotted alfalfa aphid (Nielson & Kuehl 1982). As mentioned previously, one of the reasons for the prolonged resistance of 'IR36' rice against the brown plant hopper relates to the horizontal resistance effects expressed by several minor genes. Developing horizontal resistance, however, is a long-term process that can be an extremely challenging objective for both plant breeders and entomologists. Outstanding successes in plant resistance to insects have been achieved using vertical resistance based on the sequential release of different cultivars with major genes (see Chapter 8 for a discussion of Hessian fly resistance in wheat, woolly aphid resistance in apples, and resistance to the brown plant hopper in rice).

Biotype selection is also related to the geographic extent to which resistant cultivars are planted throughout an insect's host range. Improper management practices such as elimination of alternate (weed) hosts, lack of crop rotation, or improper insecticide application may also contribute to the selection of insect biotypes on previously resistant cultivars. The occurrence of hopper biotypes in Southeast Asia is related to nearly continuous production of large-scale plantings of the same rice cultivars in several countries. Improved insect pest management techniques are now being developed and adopted to alter these practices and to avoid biotype selection.

Finally, well-defined sampling programs should be used to monitor insect populations from different geographic locations and from various host plants for biotype development. The types of methods used to determine differences vary considerably. In many insects, biotypes may be detected by the response of a group of seedling test cultivars to an insect population. In others, electrophoretic or morphometric methods may be used to detect differences between biotypes. Gonzales et al. (1978) proposed that combined electrophoretic and immunological techniques be used to delineate the genetic differences between intraspecific populations of insects with different biological characteristics. The most useful method of differentiation, however, is the one that gives the most accurate, efficient delineation of biotypes in an insect population.

Insect biotypes should be anticipated when developing plant resistance to an insect pest with a high reproductive potential. As past experiences have shown, aphid pests, with their parthenogenetic reproductive mode and high reproductive potential, are very likely to develop biotypes in response to a resistant cultivar with a single gene for antibiosis or a cultivar planted over a wide geographic range. Avoidance of large-scale monoculture and the planting of different cultivars with different genes for resistance over a wide geographic area will slow the development of insect biotypes in crop plants (see Section 8.4).

REFERENCES

Alston, F. H. and J. B. Briggs. 1977. Resistance genes in apple and biotypes of *Dysaphis devecta*. Ann. Appl. Biol. 87: 75–81.

Auclair, J. L. 1978. Biotypes of the pea aphid *Acyrthrosiphon pisum* in relation to host plants and chemically defined diets. Entomol. Exp. Appl. 24: 12–16.

Beregovoy, V. H. and K. J. Starks. 1986. Enzyme patterns in biotypes of the greenbug, *Schizaphis graminum* (Rondani) (Homoptera: Aphididae). J. Kan. Entomol. Soc. 59: 517–523.

Briggs, J. B. 1965. The distribution, abundance, and genetic relationships of four strains of the Rubus aphid (*Amphorophora rubi*) in relation to raspberry breeding. J. Hortic Sci. 49: 109–117.

Campbell, B. C. and D. L. Dreyer. 1985. Host-plant resistance of sorghum: Differential hydrolysis of sorghum pectic substance by polysaccharases of greenbug biotypes (*Schizaphis graminum*, Homoptera: Aphididae). Arch. Biochem. Physiol. 2: 203–215.

Cartier, J. J. 1959. Recognition of three biotypes of the pea aphid from southern Quebec. J. Econ. Entomol. 52: 293–294.

Cartier, J. J. 1963. Varietal resistance of peas to pea aphid biotypes under field and greenhouse conditions. J. Econ. Entomol. 56: 205–213.

Cartier, J. J. and R. H. Painter. 1956. Differential reactions of two biotypes of the corn leaf aphid to resistant and susceptible varieties, hybrids, and selections of sorghums. J. Econ. Entomol. 49: 498–508.

Chidichimo, H. O., L. B. Almaraz, B. M. Bellone, and H. O. Arriaga. 1987. Greenbug biotypes in Argentina. Plant Resist. Insects Newsl. 12: 39–40.

Claridge, M. F. and J. Den Hollander. 1982. Virulence to rice cultivars and selection for virulence in populations of the brown planthopper, *Nilaparvata lugens*. Entomol. Exp. Appl. 32: 213–221.

Claridge, M. F. and J. Den Hollander. 1983. The biotype concept and its application to insect pests of agriculture. Crop Prot. 2: 85–95.

Dahms, R. G. 1948. Comparative tolerance of small grains to greenbugs from Oklahoma and Mississippi. J. Econ. Entomol. 41: 825.

Den Hollander, J. and P. K. Pathak. 1981. The genetics of the 'biotypes' of the rice brown planthopper, *Nilaparvata lugens*. Entomol. Exp. Appl. 29: 76–86.

Dreyer, D. L. and B. C. Campbell. 1984. Degree of methylation of intercellular pectin associated with plant resistance to aphids and with induction of aphid biotypes. Experientia 40: 224.

Dunn, J. A. and D. P. H. Kempton. 1972. Resistance to attack by *Brevicornye brassicae* among plants of Brussels sprouts. Ann. Appl. Biol. 72: 1–11.

Eastop, V. F. 1973. Biotypes of aphids. Bull. Entomol. Soc. N. Z. 2: 40–51.

Eisenbach, J. and T. E. Mittler. 1987. Polymorphism of biotypes E and C of the aphid *Schizaphis graminum* (Homoptera: Aphididae) in response to different scotophases. Environ. Entomol. 16: 519–523.

Fargo, W. S., C. Inayatullah, J. A. Webster, and D. Holbert. 1986. Morphometric variation within apterous females of *Schizaphis graminum* biotypes. Res. Popul. Ecol. 28: 163–172.

Frazer, B. D. 1972. Population dynamics and recognition of biotypes in the pea aphid (Homoptera: Aphididae). Can. Entomol. 10: 1729–1733.

Gallun, R. L. 1977. The genetic basis of Hessian fly epidemics. Ann. N. Y. Acad. Sci. 287: 223–229.

Gallun, R. L. and G. S. Khush. 1980. Genetic factors affecting expression and stability of resistance. In F. G. Maxwell and P. R. Jennings (eds.). *Breeding Plants Resistant to Insects*. Wiley, New York, pp. 64–85.

Gallun, R. L. and L. P. Reitz. 1971. Wheat cultivars resistant to races of Hessian fly. US Dept. Agriculture Prod. Res. Rpt. 134, 16 pp.

Gonzales, D., G. Gordh, S. N. Thompson, and J. Adler, 1978. Biotype discrimination and its importance to biological control. In M. A. Hoy and J. J. McKelvey, Jr. (eds.). *Genetics in Relation to Insect Management. Working Papers*. Rockefeller Foundation Conference, Bellagio, Italy, March 31–April 5. Rockefeller Foundation, New York, pp. 129–136.

Guo-rui, W. C. Fu-yan, T. Lin-yong, H. Ci-Wei, and F. Bin-can. 1983. Studies on the biotypes of the brown planthopper, *Nilaparvata lugens* (Stal). Acta Entomol. Sinica 26: 154–160.

Harrington, C. D. 1943. The occurrence of physiological races of the pea aphid. J. Econ. Entomol. 36: 118–119.

Harvey, T. L. and H. L. Hackerott. 1969. Recognition of a greenbug biotype injurious to sorghum. J. Econ. Entomol. 62: 776–779.

Hatchett, J. H. 1969. Race E, sixth race of the Hessian fly, *Mayetiola destructor*, discovered in Georgia wheat fields. Ann. Entomol. Soc. Am. 62: 677–678.

Heinrichs, E. A. 1986. Perspectives and directions for the continued development of insect-resistant rice varieties. Agric. Ecosyst. Environ. 18: 9–36.

Heinrichs, E. A. and P. K. Pathak. 1981. Resistance to the rice gall midge, *Orseolia oryzae* in rice. Insect Sci. Appl. 1: 123–132.

Heinrichs, E. A. and H. R. Rapusas. 1985. Cross-virulence of *Nephotettix virescens* (Homoptera:Cicadellidae) biotypes among some rice cultivars with the same major-resistance gene. Environ. Entomol. 14: 696–700.

Hidaka, T., P. Vunsilabutr, and S. Rajamani. 1977. Geographical differentiation of the rice gall; *Orseolia oryzae* (Wood-Mason) (Diptera: Cecidomyiidae). Appl. Entomol. Zool. 12: 4–8.

Inayatullah, C., J. A. Webster, and W. S. Fargo. 1987a. Morphometric variation in the alates of greenbug (Homoptera: Aphididae) biotypes. Ann. Entomol. Soc. Am. 80: 306–311.

Inayatullah, C., W. S. Fargo, and J. A. Webster. 1987b. Use of multivariate models in differentiating greenbug (Homoptera:Aphididae) biotypes and morphs. Environ. Entomol. 16: 839–846.

Kalode, M. B., P. R. K. Viswanathan, and D. V. Seshu. 1975. Standard test to characterise host plant resistance to brown planthopper in rice. Indian J. Plant Prot. 3: 204–206.

Kalode, M. B., M. V. S. Sastry, D. J. Pophaly, and P. S. Prakasa Rao. 1976. Biotypic variation in rice gall midge, *Orseolia* (*Pachydiplosis*) *oryzae* Wood Mason Mani. J. Biol. Sci. 19: 62–65.

Keep, E. and R. L. Knight. 1967. A new gene from *Rubus occidentalis* for resistance to strains 1, 2, and 3 of Rubus aphid *Amphorophora rubi*. Euphytica 16: 209.

Kennedy, G. G., F. Gould, O. M.B. Deponti, and R. E. Stinner. 1987. Ecological, agricultural, genetic, and commercial considerations in the deployment of insect-resistant germplasm. Environ. Entomol. 16: 327–338.

Kindler, S. D. and S. M. Spomer. 1986. Biotypic status of six greenbug (Homoptera: Aphididae) isolates. Environ. Entomol. 15: 567–572.

Lammerink, J. 1968. A new biotype of cabbage aphid, *Brevicornye brassicae* L. on aphid resistant rape (*Brassica napus* L.). N. Z. J. Agric. 11: 341–344.

Lin, T. and C. Huang. 1981. Genetic studies on resistance to biotypes of the brown plant hopper (*Nilaparvata lugens*) in rice. J. Agric. Assn. China 116: 1–14.

Lowe, H. J. B. 1981. Resistance and susceptibility to colour forms of the aphid *Sitobion avenae* in spring and winter wheats (*Triticum aestivum*). Ann. Appl. Biol. 99: 87–98.

Michels, G. J., Jr. 1986. Graminaceous North American host plants of the greenbug with notes on biotypes. Southwest. Entomol. 11: 55–65.

Montllor, C. B., B. C. Campbell, and T. E. Miller. 1983. Natural and induced differences in probing behavior of two biotypes of the greenbug, *Schizaphis graminum*, in relation to resistance in sorghum. Entomol. Exp. Appl. 34: 99–106.

Niassy, A., J. D. Ryan, and D. C. Peters. 1987. Variations in feeding behavior, fecundity, and damage of biotypes B and E of *Schizaphis graminum* (Homoptera: Aphididae) on three wheat genotypes. Environ. Entomol. 16: 1163–1168.

Nielson, M. W. and H. Don. 1974. A new virulent biotype of the spotted alfalfa aphid in Arizona. J. Econ. Entomol. 67: 64–66.

Nielson, M. W. and R. O. Kuehl. 1982. Screening efficacy of spotted alfalfa aphid biotypes and genic systems for resistance in alfalfa. Environ. Entomol. 11: 989–996.

Nielson, M. W. and W. F. Lehman 1980. Breeding approaches in alfalfa. In F. G.

Maxwell and P. R. Jennings (eds.). *Breeding Plants Resistant to Insects.* Wiley, New York, pp. 279–311.

Paguia, P., M. D. Pathak, and E. A. Heinrichs. 1980. Honeydew excretion measurement techniques for determining differential feeding activity of biotypes of *Nilaparvata lugens* on rice varieties. J. Econ. Entomol. 73: 35–40.

Painter, R. H. 1930. Biological strains of Hessian fly. J. Econ. Entomol. 23: 322–326.

Painter, R. H. and M. D. Pathak. 1962. The distinguishing features and significance of the four biotypes of the corn leaf aphid *Rhopalosiphum maidis* (Fitch). Proc. XI Int. Cong. Entomol. 2: 110–115.

Pathak, P. K. and E. A. Heinrichs. 1982. Selection of biotype populations 2 and 3 of *Nilaparvata lugens* by exposure to resistant rice varieties. Environ. Entomol. 11: 85–90.

Pesho, G. R., F. V. Lieberman, and W. F. Lehman. 1960. A biotype of the spotted alfalfa aphid on alfalfa. J. Econ. Entomol. 53: 146–150.

Peters, D. C., E. A. Wood, Jr., and K. J. Starks. 1975. Insecticide resistance in selections of the greenbug. J. Econ. Entomol. 75: 339–340.

Porter, K. B., G. L. Peterson, and O. Vise. 1982. A new greenbug biotype. Crop Sci. 22: 847–850.

Puterka, G. J., D. C. Peters, D. L. Kerns, J. E. Slosser, L. Bush, D. W. Worrall, and R. W. McNew. 1988. Designation of two new greenbug (Homoptera: Aphididae) biotypes G and H. J. Econ. Entomol 81: 1754–1759.

Sato, A. and K. Sogawa. 1981. Biotypic variations in the green rice leafhopper, *Nephotettix virescens* Uhler (Homoptera: Deltocephalidae). Appl. Ent. Zool. 16: 55–57.

Saxena, R. C. and A. A. Barrion. 1983. Biotypes of the brown planthopper, *Nilaparvata lugens* (Stal). Korean J. Plant Prot. 22: 52–66.

Saxena, R. C. and L. M. Rueda. 1982. Morphological variations among three biotypes of the brown planthopper, *Nilaparvata lugens* in the Philippines. Insect Sci. Appl. 3: 193–210.

Sen Gupta, G. C. and P. W. Miles. 1975. Studies on the susceptibility of varieties of apple to the feeding of two strains of woolly aphids (Homoptera) in relation to the chemical content of the tissues of the host. Aust. J. Agric. Res. 26: 157–68.

Singh, S. R. and R. H. Painter. 1964. Effect of temperature and host plants on progeny production of four biotypes of corn leaf aphid. J. Econ. Entomol. 75: 348–350.

Singh, S. R. and E. A. Wood, Jr. 1963. Effect of temperature on fecundity of two strains of the greenbug. J. Econ. Entomol. 56: 109–110.

Sogawa, K. 1978a. Variations in gustatory response to amino. acid-sucrose solutions among biotypes of the brown planthopper. IRRN 3(5):9.

Sogawa, K. 1978b. Electrophoretic variations in esterase among biotypes of the brown planthopper. IRRN 3(5): 8–9.

Sosa, O., Jr. 1978. Biotype L, ninth biotype of the Hessian fly. J. Econ. Entomol. 71: 458–460.

Sosa, O., Jr. Biotypes J and L of the Hessian fly discovered in an Indiana wheat field. J. Econ. Entomol. 74: 180–182.

Starks, K. J. and R. L. Burton. 1972. Greenbugs: Determining biotypes, culturing, and screening for plant resistance. USDA-ARS Tech. Bull. No. 1556, 12 pp.

Starks, K. J., R. L. Burton, and O. G. Merkle. 1983. Greenbugs (Homoptera: Aphididae) plant resistance in small grains and sorghum to biotype E. J. Econ. Entomol. 76: 877–880.

Takita, T. and H. Hashim. 1985. Relationship between laboratory-developed biotypes of green leafhopper and resistant varieties of rice in Malaysia. Jap. Agric. Res. Quart. 19: 219–223.

Teetes, G. L., C. A. Schaefer, J. R. Gipson, R. C. McIntyre, and E. E. Latham. 1975. Greenbug resistance to organophosphorous insecticides on the Texas high plains. J. Econ. Entomol. 68: 214–216.

Thottappilly, G., J. H. Tsai, and J. E. Bath. 1972. Differential aphid transmission of two bean yellow mosaic virus strains and comparative transmission by biotypes and stages of the pea aphid. Ann. Entomol. Soc. Am. 65: 912–915.

Tyler, J. M., J. A. Webster, and E. L. Smith. 1985. Biotypes E greenbug resistance in wheat streak mosaic virus-resistant wheat germplasm lines. Crop Sci. 25: 686–688.

Verma, S. K., P. K. Pathak, B. N. Singh, and M. N. Lal. 1979. Indian biotypes of the brown planthopper. IRRN 4(6):7.

Webster, J. A. and C. Inayatullah. 1985. Aphid biotypes in relation to host plant resistance: a selected bibliography. Southwest. Entomol. 10: 116–125.

Whalon, M. E., G. Wilde, and H. Feese. 1973. A new corn leaf aphid biotype and its effect on some cereal and small grains. J. Econ. Entomol. 66: 570–571.

Wilde, G. and H. Feese. 1973. A new corn leaf aphid biotype and its effect on some cereal and small grains. J. Econ. Entomol. 66: 570–571.

Wood, E. A., Jr. 1961. Biological studies of a new greenbug biotype. J. Econ. Entomol. 54: 1171–1173.

Woottipreecha, S. 1971. Studies of genetic synthesis of new races of Hessian fly, *Mayetiola destructor* (Say). Ph.D. thesis, Purdue University, West Lafayette, IN. 187 pp.

SECTION III

How Can Plant Resistance to Insects Be Utilized?

10 Use of Plant Resistance in Insect Pest Management Systems

10.1 SIGNIFICANCE OF DIFFERENT CATEGORIES OF PLANT RESISTANCE IN INSECT PEST MANAGEMENT SYSTEMS

The type of insect resistance deployed in an integrated insect pest management (IPM) system has a direct influence on the stability and ultimate success of an insect-resistant cultivar. Different categories of resistance have different degrees of effectiveness in an IPM system, depending on the movement and host preferences of the pest insect (s). Simulation modeling studies by Kennedy et al. (1987) indicate that low levels of antixenosis, antibiosis, and tolerance can be effective in controlling resident insect pests that invade plantings early in the development of a crop and increase gradually during the growing season. Because the movement of resident insects is inherently limited, antixenosis may be adequate to reduce populations of this type of insect in an IPM system. The results of Alvarado-Rodriguez et al. (1986) indicate that feeding antixenosis resistance in common bean cultivars to the lygus bug, *Lygus hesperus* Knight, should be effective against both migratory and endemic lygus bug populations. Leszczynski (1987) has drawn similar conclusions concerning antixenosis in European wheat to the grain aphid, *Sitobion avenae*. Antixenosis due to the presence of awns on grain heads may be more beneficial than antibiosis resistance to migrating populations of the aphid.

Cropping systems attacked by highly mobile pests however, require different categories and levels of resistance for their effective management. In additional research, Kennedy et al. (1987) used simulation modeling to compare the effects of resistance in tomato and soybean on the corn earworm, *Heliothis zea* Boddie. These results (Figure 10.1) indicate that high mortality in early instar larvae from tomato antibiosis or oviposition antixenosis will reduce earworm populations, but insecticidal control is still necessary. However, because soybean plants can withstand greater defoliation than tomato plants, corn earworm populations reach economically damaging levels less often on soybeans than on tomatoes. For this reason, low or moderate levels of antibiosis or antixenosis should be useful in reducing corn earworm infestations on soybean.

The use of tolerant cultivars in IPM systems also offers several advantages. Insect population levels are not diminished from exposure to tolerant plants, as they are on plants exhibiting antibiosis and antixenosis, but their gene pools remain diluted, because the selection pressure placed on them by high levels of antibiosis is reduced or absent. Thus the potential for development of resistance-breaking biotypes (see Chapter 9) is greatly diminished through the use of tolerant cultivars (Teetes 1980). Some maize cultivars tolerant of damage by the corn earworm, *Heliothis zea* (Boddie), and the European corn borer, *Ostrinia nubilalis* Hubner, actually harbor larger larval populations than susceptible cultivars, due presumably to their increased biomass (Wiseman et al. 1972, Hudon et al. 1979), but this does not decrease their effectiveness in providing greater yields than susceptible cultivars.

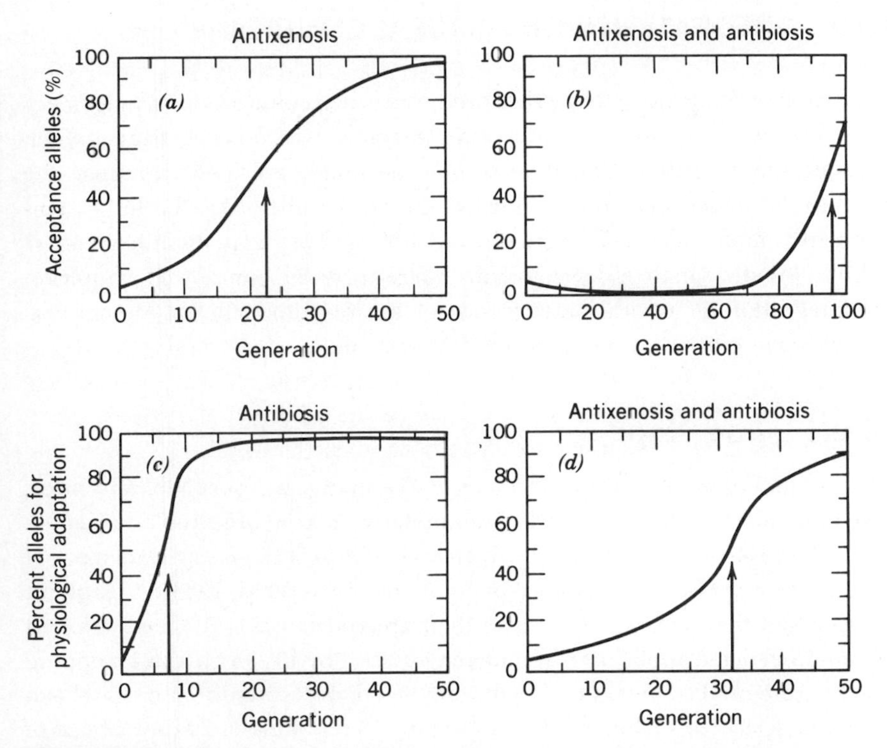

FIGURE 10.1 Durability of antibiosis and antixenosis resistance of silking-stage corn to the corn earworm, *Heliothis zea* Boddie, when the two types of resistance are used singly (A and C) or together in a single cultivar (B and D). C and D show the increase in the frequency of alleles for *H. zea* adaptation to antibiosis. A and B show the increase in frequency of alleles in *H. zea* for accepting the antixenotic corn. (From Kennedy et al. 1987. Reprinted with permission from Environ. Entomol. 16: 327–338. Copyright 1987, Entomological Society of America.)

Tolerance also enhances the effects of biological control agents in crop protection systems. Tolerant cultivars do not expose beneficial insects to the adverse effets of plant morphological or allelochemical factors in cultivars that exhibit antibiosis or antixenosis (see Chapters 2 and 3). From the perspective of the total effect of the resistant plant on the insect population, cultivars with tolerance require less antixenosis or antibiosis than cultivars without tolerance. Brewer et al. (1986) suggest that production of alfalfa cultivars possessing tolerance, antibiosis, and antixenosis resistance to the potato leafhopper, *Empoasca fabae* (Harris), may provide stable leafhopper control by raising the leafhopper economic injury level.

10.2 INTEGRATION WITH CULTURAL CONTROL

Variations in cultural practices in production agriculture is a management tactic that was first used by Chinese rice farmers over 2000 years ago, when adjusted planting dates and crop stubble burning were used to reduce the buildup of insect pest populations (Flint and vandenBosch 1981). Undoubtedly, many other early agricultural systems implemented similar types of cultural practices in their development. There are few documented examples of the integration of cultural practices with plant resistance, but many cultural practices are effective when used in IPM systems.

10.2.1 Trap Crops

The combination of a trap crop, the early maturing, okra leaved cotton breeding line (La 1363 Lsne), and a nonpreferred cotton breeding line (La 81-560FN) resistant to the boll weevil, *Anthomonous grandis grandis* Boheman, is effective in suppressing boll weevil populations (Burris et al. 1983). Treatment of weevils in the concentrated area of the trap crop causes a 20% reduction in overall insecticide application and increases yields by 14–33%. Trap cropping also has potential for integration with the growth of resistant rice and soybean cultivars. Rice fields planted adjacent to a trap crop planted 20 days ahead of the main crop attract more brown plant hoppers, *Nilaparvata lugens* (Stal), yield significantly more, and preserve more natural enemies than control fields without trap crops (Saxena 1982). Cowpea trap crops planted prior and adjacent to soybean plantings are also effective in concentrating populations of the southern green stink bug, *Nezara viridula* (L.), where they can be controlled with insecticides, allowing fewer or no insecticidal treatments of the main crop (Newsom et al. 1975).

10.2.2 Early Maturing Crops

Planting the rice crop in a manner that avoids the peak pest insect population generally reduces insect pest damage. Heinrichs et al. (1986a) demonstrated that brown plant hopper populations and hopper:predator ratios on very early and early maturing rice cultivars are significantly lower than those on mid-season maturing cultivars. These results indicate that the incorporation of brown plant hopper resistance into early maturing cultivars would serve to increase rice crop protection in the event of hopper biotype development (Fig. 10.2).

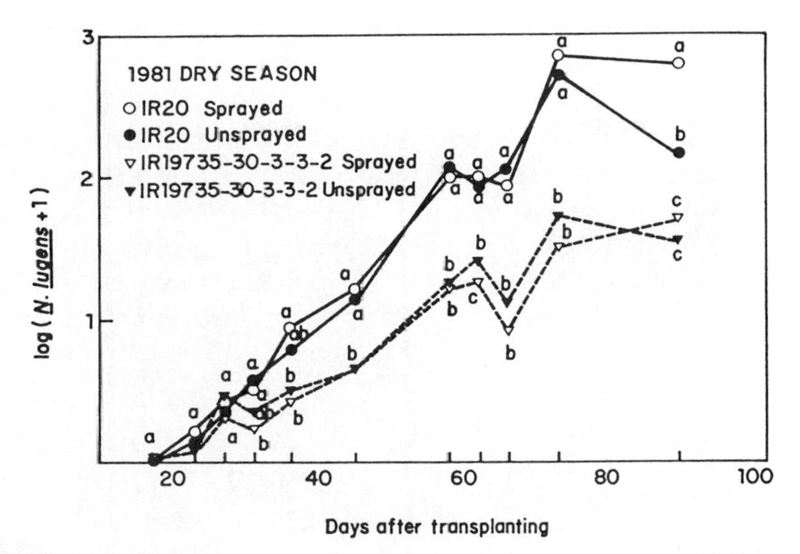

FIGURE 10.2 Brown plant hopper, *Nilaparvata lugens* Stal, populations in insecticide treated and untreated plots of very early maturing (IR 19735-30-3-3-2) and mid-season maturing (IR20) rice cultivars. At each date, points having different letters (a-c) are different at $p < .05$. (From Heinrichs et al. 1986a. Reprinted with permission from Environ. Entomol. 15: 93–95. Copyright 1986, Entomological Society of America.)

Similarly, early maturing cotton cultivars that contain the nectaryless trait (lack of extrafloral nectaries) have yields in untreated plots that are only slightly less than those of insecticide treated plots (Bailey et al. 1980). The improved cotton cultivar 'Gumbo 500' matures earlier than other adapted cultivars, eliminating the need for one to two insecticide applications (Jones et al. 1981).

10.3 INTEGRATION WITH CHEMICAL CONTROL

10.3.1 Cotton

Two morphological characteristics of improved cotton cultivars open up the plant canopy and increase the efficiency of insecticides applied to control cotton insect pests. The frego or open bract condition of cotton buds (squares) (Fig. 10.3) allows increased penetration of insecticides into the areas around each boll and increases mortality in treated infestations of the boll weevil (Parrott et al. 1973). Combinations of the okra leaf (open leaf) (Fig. 10.4) and frego

FIGURE 10.3 Frego bract (left) and normal bract (right) cotton flower buds. (Courtesy J. E. Jones, Louisiana State University.)

bract characters improve insecticide efficiency by increasing coverage on all plant parts (Jones et al. 1981, 1986). The effects of early maturity and resistance in cotton to the pink bollworm, *Pectinophora gossypiella* Saunders, together exceed the benefits of insecticide applications. George and Wilson (1983) found no significant reduction in yield between insecticide treated and untreated plantings of the resistant line 'AET-5,' even though the resistant line had a greater level of bollworm infestation. In the same experiment, a commercial cotton cultivar sustained a 16% yield reduction when not treated with insecticide.

10.3.2 Cereals

The planting of sorghum cultivars with moderate resistance to the sorghum midge has raised economic thresholds for insecticidal control of the sorghum midge, *Contarinia sorghicola* (Coquillett). Untreated midge-resistant sorghum hybrids yield more than treated susceptible hybrids, and at moderate and high midge densities, resistant respond more efficiently to insecticide application in terms of net crop yield and value (Teetes et al. 1986).

Maize hybrids resistant to the corn earworm require less insecticide than susceptible hybrids to achieve equivalent levels of corn earworm control. One application of insecticide at one-half of the normal rate on the resistant maize

FIGURE 10.4 (*a*) Okra and (*b*) normal cotton leaves. (Courtesy J. E. Jones, Louisiana State University.)

hybrid '471-U6X81-1' controls corn earworm larvae at a level equal to seven applications of the same insecticide at the normal rate on a susceptible hybrid (Wiseman et al. 1975). However, application of insecticides to maize hybrids with intermediate and high levels of resistance to the European corn borer, is of little benefit in reducing borer damage in the field (Robinson et al. 1978).

The susceptibility of the brown plant hopper and the white-backed plant hopper, *Sogatella furcifera* (Horvath), to insecticides is increased when hoppers are reared on only moderately hopper-resistant rice cultivars (Heinrichs et al. 1984). Application of insecticide to brown plant hopper-resistant rice cultivars increases hopper mortality for prolonged periods after transplanting and decreases insecticide usage. As demonstrated by Heinrichs et al. (1986b), gross profit and net gain on investment are also greater when insect-resistant rice cultivars are grown than when susceptible rice cultivars are produced. Insecticides also significantly enhance the yield of insect resistant rice cultivars. Carbofuran treatment of 'IR26,' a cultivar with resistance to brown plant hopper biotype 1, controls the rice whorl maggot, *Hydrellia philippina*, and increases rice yields by 15% (Heinrichs et al. 1979).

In some instances, rice resistance is great enough that insecticides are of no practical value. Kalode (1980) indicated that cultivars with resistance to the gall midge, *Orseolia oryzae* Wood-Mason, yielded similarly, regardless of how insecticides were applied (Table 10.1). Though the same trend was observed for insecticide treatments to the susceptible cultivar, yields were nearly one-half that of the resistant cultivar. Similar studies by Reissig et al. (1981) in the Philippines indicate that applications of insecticide to 'IR36,' a multiple insect-

TABLE 10.1 Yields of Resistant (R) 'Shakti' and Susceptible (S) 'Jaya' Rice Cultivars after Managed and Unmanaged Insecticide Control of the Rice Gall Midge, *Orseolia oryzae*, in Sambalpur, Cuttack, and Bhubaneswar, India 1977–1978[a]

	Shakti(R)		Jaya(S)	
Insecticide Treatment Method	Silver Shoots (%)	Yield (t/ha)	Silver Shoots (%)	Yield (t/ha)
Surveillance-based	2.3	3.5	12.7	2.2
Prophylactic	1.0	3.3	9.8	2.3
None	2.7	3.3	21.7	1.8

[a]From Kalode (1980).

TABLE 10.2 Mortality of Brown Plant Hoppers, *Nilaparvata lugens*, reared on Resistant (R), Moderately Resistant (MR), and Susceptible (S) Rice Cultivars after Treatment with Various Insecticides, Los Banos, Philippines, 1980

	Mean 24-hr Mortality		
Insecticide	Taichung Native 1 (S)	ASD7 (MR)	Sinna Sivappu (R)
Carbofuran 12F	64b*	94a	95a
Carbosulfan 20EC	66b	93a	94a
BPMC 50EC	46b	79a	90a
Control	10b	18a	23a

From Heinrichs et al. (1984). Adapted with permission of Environ. Entomol. 13: 455–458. Copyright 1984, Entomological Society of America.
*$p = .05$

and disease-resistant cultivar, for brown plant hopper and green leafhopper control do not increase yields and actually decrease farmers' net profits. Heinrichs et al. (1984) has shown that brown plant hopper mortality due to insecticide application is not different between hoppers reared on resistant and moderately resistant cultivars (Table 10.2). Thus the use of moderate resistance is of real value in reducing the degree of selection pressure causing brown plant hopper biotype development.

10.3.3 Legumes

Chalfant (1965) noted a positive, synergistic interaction between bean cultivars resistant to the potato leafhopper, *Empoasca fabae* Harris, and toxicity of the insecticide carbaryl, when both were used in field evaluations for suppression of potato leafhopper populations. Campbell and Wynne (1985) determined that insecticide applications for control of the southern corn rootworm, *Diabrotica undecimpunctata howardi* LeConte, on the resistant peanut cultivar 'NC6' may be reduced by 80% of the normally applied amount of insecticide (Fig. 10.5). Similarly, applications for control of thrips and potato leafhopper may be reduced by about 60%. These reductions in insecticide volumes significantly reduce peanut production costs.

As in rice, legume resistance to insects may replace insecticidal control completely in a management system. Research by Cuthbert and Fery (1979) in

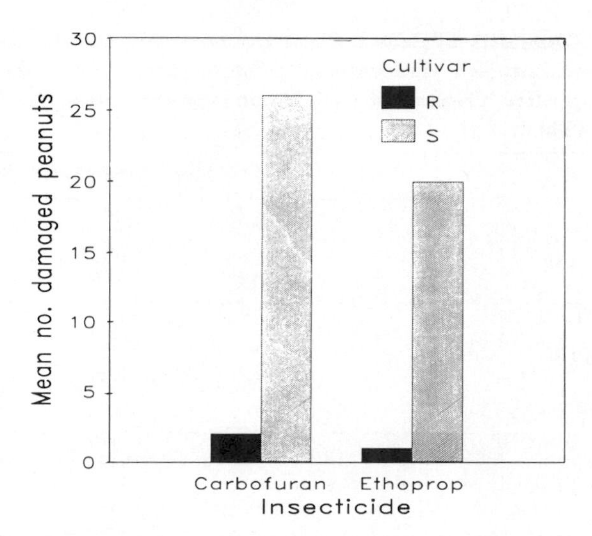

FIGURE 10.5 Reduction in populations of the southern corn rootworm, *Diabrotica undecimpunctata howardi* LeConte, by insecticides and the resistant cultivar 'NC6'. R = resistant; S = susceptible. (From Campbell and Wynne 1985. Adapted with permission from J. Econ. Entomol. 78: 113–116. Copyright 1985, Entomological Society of America.)

Georgia indicates that resistance in southern pea to the cowpea curculio, *Chalcodermus aeneus* Boheman, is actually more effective in reducing curculio damage than applications of insecticide. Resistance in sweet potato is also more effective or as effective as insecticide, depending on the use of a highly or moderately resistant cultivar, in reducing injury by an insect complex consisting of the southern potato wireworm, *Conoderus falli* Lane, the banded cucumber beetle, *Diabrotica balteata* LeConte, the spotted cucumber beetle, *Diabrotica undecimpunctata howardi* LeConte, the elongate flea beetle, *Systena elongata* (F.), the white grub, *Plectris aliena* Chapin, and the sweet potato beetle, *Cylas formicarius elegantulus* (Summers) (Cuthbert & Jones 1978).

Larvae of the soybean looper, *Pseudoplusia includens* (Walker), feeding on foliage of the insect-resistant soybean breeding line 'ED73-371' are more susceptible to poisoning by methyl parathion insecticide than larvae feeding on the susceptible cultivar 'Bragg,' in both laboratory and field assays (Kea et al. 1978). In the laboratory, the susceptibility of soybean looper larvae and velvetbean caterpillar, *Anticarsia gemmatalis* (Hubner), larvae to poisoning by the insecticides fenvalerate and acephate is also enhanced by prior consumption of foliage of the insect-resistant 'PI (plant introduction) 227687' (Rose et al. 1988).

10.3.4 Bacterial Insecticides

Several studies have documented the combined beneficial effects of the bacterial insecticide, *Bacillus thuringiensis* (*B.t.*), with resistant cultivars and conventional insecticides on infestations of lepidoterous larvae. Combined applications of *B.t.* and the insecticide chlordimeform significantly reduce larval populations of the cabbage looper, *Trichoplusia ni* (Hubner) and the imported cabbageworm, *Artogeia rapae* (L.). Some cultivars have fewer uninjured plant parts after treatment, indicating the complementary effect of plant resistance, a microbial insectide, and a conventional insecticide (Creighton et al. 1975). Complementary effects also exist between cabbage resistance and *B.t.* alone for control of larval feeding damage by the cabbage looper, imported cabbageworm, and diamondback moth, *Plutella xylostella* (L.) (Creighton et al. 1981).

Combinations of *B.t.* and the fungus *Nomuraea rileyi* increase mortality of larvae of the bollworm, *Heliothis zea* (Boddie), on an insect-resistant soybean cultivar by approximately 40% (Bell 1978). Larval mortality on untreated plants is about 60% owing to the effects of resistance alone, but increases to 100% when the microbial insecticides are applied. Similar results were noted by Kea et al. (1978) in experiments to determine the combined effects of the multiple insect-resistant cultivar ED73-371 and *B.t.* on field populations of the bollworm. Application of *B.t.* to the resistant cultivar gave greater control of bollworms than when the pathogen was applied to bollworms infesting the susceptible cultivar 'Bragg.'

10.3.5 Possible Negative Interactions of Plant Resistance to Insects with Insecticides

In spite of many positive interactions between plant resistance and insecticides, there are possibilities for negative interactions between the two tactics. Insects possess a variety of detoxification mechanisms that allow them to survive on plants containing defensive allelochemicals. Increases in the activity of digestive enzymes such as mixed function oxidases, glutathione s-transferases, and hydrolases can be induced by a wide range of allelochemicals (Brattsten 1979, Yu 1982, Yu & Hsu 1985). Insects with such induced enzyme systems may be more tolerant of the toxic effects of insecticides normally detoxified by these same enzymes.

Several studies have investigated the effects of selected allelochemicals from different host plants on insect metabolism. These include the allelochemical sinigrin (Yu 1983) and xanthotoxin (Yu 1985). However, information about the interactions that exist between allelochemicals controlling the resistance of crop plants and the metabolic capability of the associated pest insect is just beginning to be developed.

As indicated in Section 3.3, 2-tridecanone from *Lycopersicon hirsutum* f. *glabratum* increases the ability of the tobacco hornworm, *Manduca sexta* (L.), to detoxify the insecticide carbaryl (Kennedy 1984). In a similar manner, the allelochemical gossypol (from insect-resistant cultivars of cotton) increases the activity of the *N*-demethylase enzyme in the cotton leafworm, *Spodoptera littoralis* (El-Sebae et al. 1981). In both cases, the potential exists for a negative interaction between allelochemically based resistance to pest Lepiodoptera and conventional insecticidal control measures for these pests.

Potential incompatability between plant resistance and microbial insecticides also exists. A great deal of emphasis is currently being placed on incorporating the genes that encode for the protein toxins from *B.t.* into crop plants (see Section 6.3.3). In this case the insecticide serves as the basis of the resistance. However, as with hundreds of other cases of insecticide resistance, the metabolic ability of insects to detoxify *B.t.* has also been demonstrated. McGaughey and Beeman (1988) demonstrated that resistance to *B.t.* in the Indian mealmoth, *Plodia interpunctella* (Hubner) occurs in as little as two to three generations.

These studies indicate the need to develop crop cultivars with moderate levels of microbial toxin resistance, in order to avoid insect biotypes that will be easily induced to overcome the resistance. These studies also indicate the need for much more research on the outcomes of the interactions between insecticides and plant resistance. This research should address the effects of both insecticides and plant allelochemicals on insect detoxification systems in order to avoid development of resistant crop plants that are incompatible with insecticides.

10.4 INTEGRATION WITH BIOLOGICAL CONTROL

10.4.1 Viruses and Fungi

The interactions of viruses and fungi with insect-resistant cultivars are largely unexplored. The results of Hamm and Wiseman (1986) however, confirm the existence of a synergistic interaction between maize cultivars with resistance to leaf feeding by the fall armyworm, *Spodoptera frugiperda* (J. E. Smith), and the nuclear polyhedrosis virus (NPV). Larvae feeding on an artificial diet containing freeze-dried silks from fall armyworm-resistant inbred lines or on the intact silks of field grown plants are more susceptible to infection and mortality from NPV than larvae fed similarly with silks from susceptible inbred lines. Allelochemicals mediating plant resistance may adversely affect the synergism of resistance with the effects of NPV. The phenolics rutin and chlorogenic acid, which may mediate tomato resistance to insects, inhibit the rate of infection of NPV in corn earworm larvae and extend the survival time of infected larvae (Felton et al. 1987).

No clearly defined synergism between entomophagous fungi and insect-resistant cultivars is known to exist. However, Ramoska and Todd (1985) noted that chinch bugs fed corn and sorghum after inoculation with the fungus *Beauveria bassiana* suffer lower mortality than bugs fed wheat, barley, water, or artificial diet. The spread of fungal spores from cadavers of infected individuals also progresses at a lower rate from bugs fed corn and sorghum. These results indicate that fungal inhibitors produced by some plants may protect phytophagous insects from infection by fungal pathogens, whereas the lack of inhibitors in other plants may allow fungal invasion of insects feeding on these plants.

10.4.2 Predators and Parasites

Research in integrated pest management is beginning to establish the basis and degree of beneficial interactions of plant resistance with insect predators and parasites. Predators and parasites effectively complement the effects of insect-resistant cultivars of alfalfa, cotton, maize, potato, rice, sorghum, soybean, and wheat. As indicated by Bergman and Tingey (1979), the activities of insect parasites and predators are generally not adversely affected by genetic plant resistance, even though the population levels of host insects are reduced. However, several recent studies have documented more specifically the effects that physical and chemical changes in insect-resistant plants have on beneficial insects. Boethel and Eikenbary (1986) produced a more recent comprehensive review of both the fundamental and applied aspects of plant resistance interaction with insect natural enemies.

One of the first documented examples of the positive interaction of plant resistance and biological control was by Starks et al. (1972), who determined that the effects of barley and sorghum cultivars resistant to the greenbug are complemented by the activity of the parasite *Lysiphlebus testaceipes* (Cresson). (Fig. 10.6). The effectiveness of *L. testaceipes* remains unaffected, even after the development of additional greenbug biotypes (Salto et al. 1983). In a related cereal crop, predation of the brown plant hopper and the green leafhopper,

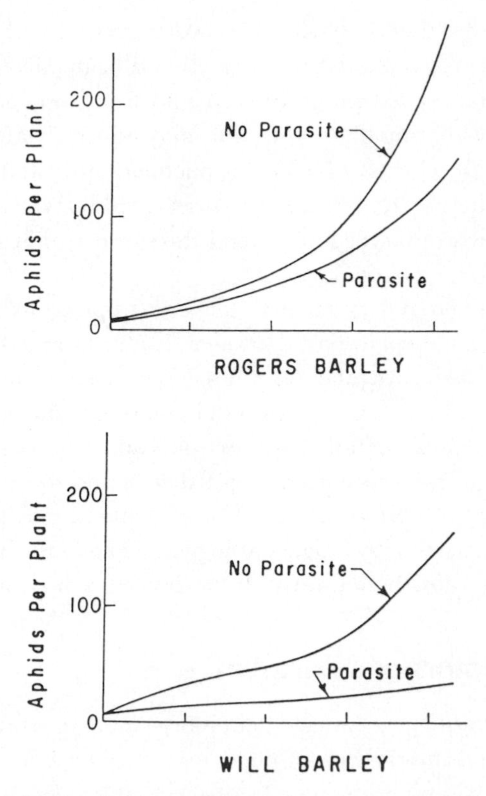

FIGURE 10.6 Population development (6 aphids/plant initially) of the greenbug, *Schizaphis graminum* (Rondani), on resistant ('Will') and susceptible ('Rogers') barley plants with and without parasitization by *Lysiphlebus testaceips* (Cresson). (From Starks et al. 1972. Reprinted with permission from Ann. Entomol. Soc. Am. 65: 650–655. Copyright 1972, Entomological Society of America.)

Nephotettix virescens (Distant), by the spider, *Lycosa pseudoannulata* Boes. et Str., or the mirid bug *Cyrtorhinus lividipennis* Reuter, is higher on hopper-resistant rice cultivars. This combination of moderate plant resistance and predation keeps hopper population levels below the economic threshold on resistant and moderately resistant rice cultivars (Kartohardjono and Heinrichs 1984, Myint et al. 1986) (Table 10.3).

The maize hybrids 'Dixie 18' and '471–U6 × 81 – 1,' which have tolerance to the corn earworm, maintain high levels of the predator *Orious insidiosus* Say from silking to 1 week after silking. These predator populations contribute to a greater suppression of corn earworm larval populations on the resistant hybrid

TABLE 10.3 Percent Mortality of Brown Plant Hoppers, *Nilaparvata lugens*, due to the Combined Effects of Moderately Resistant (MR) and Susceptible (S) Rice Cultivars and Predators, Los Banos, Philippines, 1981

	Triveni(MR)		Taichung Native(S)	
Predator	Without Predators	With Predators	Without Predators	With Predators
Lycosa pseudoannulata	26.9bc*	46.6a	14.8d	36.0ab
Cyrtohinus lividipennis	8.3b	34.2a	0.0c	10.0b

From Kartohadjono and Heinrichs (1984). Adapted with permission from Environ. Entomol. 13: 359–365. Copyright 1984, Entomological Society of America.
*$p = .05$.

than on susceptible hybrids (Wiseman et al. 1976). The protozoan parasite *Nosema pyrausta* and maize cultivars resistant to leaf and sheath-collar feeding by the European corn borer interact to reduce borer populations significantly (Lynch & Lewis 1976, Lewis & Lynch 1976). Isenhour and Wiseman (1987) found a synergistic interaction between genotypes of maize resistant to the fall armyworm and the armyworm parasite, *Campoletis sonorensis* (Cameron). Parasitism results in further reductions in armyworm larval weights over those caused by consumption of resistant foliage alone (Fig. 10.7), and the resistance has no adverse affects on parasite development.

One of the same genetic traits in improved cotton cultivars that complement the effectiveness of insecticides also functions synergistically with naturally occurring predators and parasites to suppress boll weevil populations. Larvae infesting cotton cultivars with the frego bract trait sustain approximately 40% greater parasitism by the parasitic wasp, *Bracon melitor* Say than larvae infesting bolls on cultivars with the normal closed bract condition (McGovern & Cross 1976).

Bergman and Tingey (1979) indicated that high levels of plant pubescence in insect-resistant cultivars could have detrimental effects on predators and parasites. Galbrous (nonpubescent) cottons alter bollworm oviposition behavior, resulting in reduced bollworm populations. However, the yields of cotton cultivars with the glabrous trait are also reduced (Lukefahr 1971). Treacy et al. (1985) studied the effects of the parasite *Trichogramma pretiosum* (Riley) and the predator *Chrysopa rufilabris* (Burmeister) on bollworm larvae feeding on glabrous-leaved, pubsescent, and densely pubescent cotton cultivars. The effects of both beneficial insects are reduced with increasing degrees of cotton leaf pubescence. Thus although glabrous-leaved cotton cultivars

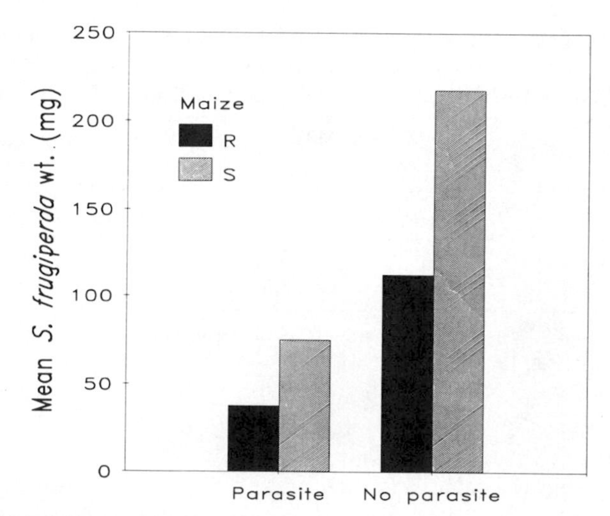

FIGURE 10.7 Effect of parasitism by *Compoletis sonorensis* on the weights of fall armyworm, *Spodoptera frugiperda* (J. E. Smith), larvae feeding on resistant (black bar) and susceptible (gray bar) maize foliage. (From Isenhour and Wiseman 1987. Adapted with permission from Environ. Entomol. 16: 1181–1184. Copyright 1987, Entomological Society of America.)

reduce bollworm populations, bollworm predators and parasites are adversely affected. Schuster and Calderon (1986) observed that numbers of predators actually increase on pubescent cottons, but that predator searching efficiency decreases.

The density of glandular trichomes on the stems of tomato plants also affects the ability of the predatory mite, *Phytoseiulus persimilis* Athiaus-Henriot, to control the phytophagous two-spotted spider mite, Tetranychus urticae Koch (van Haren et al. 1987). Spider mites are also affected by the trichomes, but to a lesser extent, owing to the ability to reach plant tissues by descent on silk threads and to spin silken mats over areas of glandular trichomes. Tomato cultivars without trichomes allow leaf to leaf dispersal of the predatory mites, enabling them to effectively control the pest mites.

Obrycki and Tauber (1984) and Obrycki et al. (1983) determined that moderate levels of glandular pubescence in potato hybrids derived from the wild potato, *Solanum berthaultii*, integrate effectively with predators and parasites in the management of green peach aphid, *Myzus persicae* (Sulzer), potato aphid, *Macrosiphon euphorbiae* (Thomas), and pea aphid, *Acyrthosiphon pisum* (Harris), populations infesting potatoes. Lampert et al. (1983) determined that resistance in small grain cultivars to the cereal leaf beetle, *Oulema melanopus* (L.), based on increased leaf pubescence has very little negative effect on cereal leaf beetle parasites.

Suppression of beneficial arthropod populations may also result from the effects of nonspecific allelochemicals in resistant plant cultivars. The first example of this interaction was observed in the toxicity of α-tomatine, an alkaloid from resistant tomato cultivars, to *Hyposoter exiguae* (Viereck), an endoparasite of the tomato fruitworm, *Heliothis zea* (Boddie) (Campbell & Duffey 1979).

The lack of adequate nutrients may also be detrimental to populations of beneficial arthropods. Nectaryless cotton genotypes suppress oviposition and ensuing infestations of cotton by lepidopterous larvae, but the removal of the extrafloral nectaries, a high energy insect food source, from the cotton plant also reduces the effectiveness of several predators and parasites (Treacy et al. 1987, Schuster et al. 1976).

In addition to differences between resistant and susceptible cultivars in nutrient chemical composition, toxin content, and morphological factors, differences in the volatile chemical differences between resistant and susceptible cultivars (Chapter 2) may also affect natural enemies (Price 1986). If resistant cultivars contain volatiles that attract beneficial insects, then greater numbers of parasites will be drawn to the plant. Conversely, compounds repellent to beneficial insects will drive them from the plant and their prey (van Emden 1986). Associational resistance (Chapter 1) related to the volatile content of companion crops grown with resistant cultivars may also have positive (attractant) or negative (repellent) effects on beneficial insects seeking prey on the target crop plant (Price 1986).

High levels of resistance in soybeans to several insects are also detrimental to parasites. Parasitization of soybean looper larvae by the hymenopterous parasite *Microplitis demolitor* Wilkinson and the resistance of soybean PI227687 reduce looper larval consumption of soybean foliage more than resistance alone. However, the high level of allelochemically based antibiosis resistance in PI227687 (Reynolds et al. 1985) causes high mortality of soybean looper larvae and decreases *M. demolitor* survival in later generations (Yanes & Boethel 1983) (Fig. 10.8).

Development of the parasite *Microplitis croceipes* Cresson in bollworm larvae fed foliage of PI227687 and PI229358 is also greatly delayed (Powell & Lambert 1984). More successful development occurs among larvae fed the moderately resistant plant introduction PI171451 and the susceptible cultivar 'Davis.' Consumption of PI171451 and 'Davis' foliage by bollworm larvae is

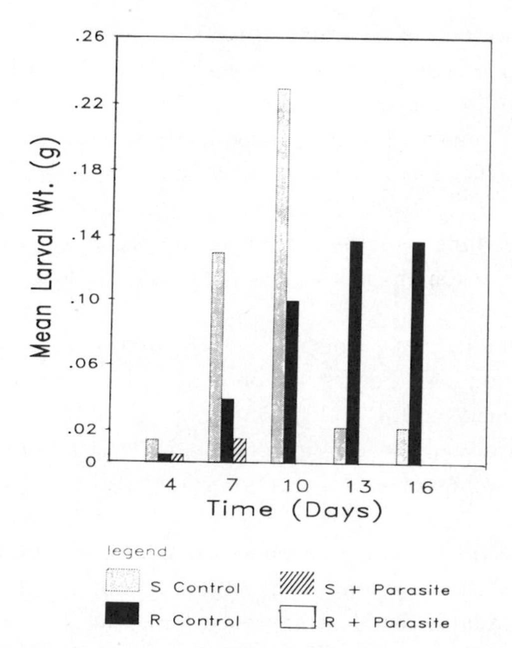

FIGURE 10.8 Weights of soybean looper, *Pseudoplusia includens* Walker, larvae fed resistant (R) or susceptible (S) soybean foliage after parasitization by *Microplitis demolitor* Wilkinson. (From Yanes and Boethel 1983. Reprinted with permission from Environ. Entomol. 12: 1270–1274. Copyright 1983, Entomological Society of America.)

greatly reduced because of parasitism. A similar tritrophic level interaction occurs between soybean looper larvae, resistant soybean cultivars, and the hymenopterous parasite *Copidosoma truncatellum* (Dalman) (Orr & Boethel 1985, Beach & Todd 1986).

The effect of parasitism on the Mexican beam beetle, *Epilachna varivestis* by *Pediobius foveolatus* Crawford, a larval endoparasite, is greater on beetles fed the resistant soybean cultivar 'Cutler 71' than on those fed the susceptible cultivar 'Bonus' (Kauffman & Flanders 1985). The higher the level of soybean resistance, the greater the negative impact on parasite survival and development. Orr et al. (1985) demonstrated that emergence, fecundity, and progeny production of the parasite *Telenomus chloropus* Thompson are reduced when parasites develop on eggs of the southern green stinkbug that are fed pods of the resistant soybean cultivar PI171444 compared to stinkbugs fed pods of the susceptible cultivar 'Davis.'

Orr and Boethel (1985) evaluated a four trophic level interaction between

PI227687 foliage, soybean looper larvae, the pentatomid predator *Podisus maculiventris* (Say), and the pentatomid parasite *Teleonomus podisi* Ashmead. As in tritrophic level interactions, predator development was adversely affected after consumption of looper larvae fed resistant soybean foliage. The development of *T. podisi.* after development on *P. maculiventris* reared on looper larvae fed PI227687 foliage was unaffected, but the reproductive potential of the parasite did decrease. Thus the antibiosis resistance expressed by soybean PI227687 is expressed through four trophic levels.

10.4.3 Optimizing Biological Control and Plant Resistance in Insect Pest Management

It is difficult to develop a general theory describing the optimum level of plant resistance and pest insect predation or parasitism. Though the literature discussed represents a diversity of crop plants (Gramineae, Leguminosae, Malvaceae, Solanaceae) and pest arthropods (Acarina, Hemiptera, Homoptera, Lepidoptera), too few studies have been conducted to develop a general model of the interaction of plant resistance and biological control. However, in the instances where plant resistance has been shown to have negative effects, beneficial insects were exposed to high levels of allelochemical or morphological resistance. Where a moderate degree of resistance is employed, positive, beneficial effects on natural enemies have resulted in reduced populations of the pest insect.

In addition to plant factors affecting the success or failure of plant resistance and biological control, ecological factors directly related to beneficial insects themselves may also determine the outcome of the interaction of plant, pest insect, and natural enemy. These factors may include the type and degree of density dependence of the natural enemy of the pest insect, alternate food preferences of the natural enemy, and other causes of pest mortality, such as entomopathogens. In a unique study to investigate some of these factors, van Emden (1986) modeled natural enemy effectiveness on moderately resistant cultivars of different crops. Both natural enemy attraction to prey and the amount of time required to find and handle prey were actually greater than expected in relation to the reduced prey density on the resistant cultivars.

From the perspective of integrated pest management, the ideal interaction of plant resistance and biological control is one that reduces pest insect populations so that beneficial insects restrict pest population levels to slightly below the economic injury level on a crop plant.

10.5 CONCLUSIONS

Plant resistance to insects has been successfully integrated with all other types of insect pest management tactics in many of the world's major food and fiber crops. However, in order to be of greatest utility, insect-resistant cultivars must fit efficiently into existing IPM systems based on previously developed cultural practices and insecticide use patterns. The fit of an insect-resistant cultivar into the management system is also enhanced if it exhibits resistance to several insect species, as opposed to a single species.

Multiple species resistance is especially beneficial, because it may result in a greater reduction in the total amount of insecticides applied to the system than that resulting from the production of a cultivar with resistance to only one insect. Multi-species resistance is also helpful in avoiding the emergence of a secondary insect pest species as a primary pest. Secondary pests often take on primary pest status as a result of reductions in the amounts of insecticides applied for control of the original primary pest insect.

In spite of these obvious advantages, multiple insect resistance is often difficult to develop. The difficulties involved in the development of cotton cultivars with multiple insect resistance discussed by Jenkins (1980) illustrate this problem. Three major cotton morphological plant factor characters, smooth leaf, frego bract, and lack of nectaries (see Section 10.2.2), have varying effects on three major cotton insect pests. These include the boll weevil, tarnished plant bug, *Lygus lineolaris* (Palisot de Beauvois), and *Heliothis* spp. larvae. The smooth leaf character reduces infestation and boll damage by *Heliothis* spp. larvae but increases tarnished plant bug feeding injury. The frego bract character reduces boll weevil oviposition but increases susceptibility to tarnished plant bug feeding damage. The absence of extrafloral nectaries reduces damage resulting from both *Heliothis* spp. larvae and tarnished plant bugs, but has no effect on the boll weevil. In addition, nectaries also serve as sources of food for several parasites and predators that are important as natural population regulating factors (see Section 10.4.2). In spite of these obstacles, multiple adversity resistant cottons have been developed that contain resistance to the bollworm, tarnished plant bug, boll weevil, and cotton fleahopper, *Pseudatomoscelis seriatus* (Reuter) (Bird 1982). Multiple insect species resistance has also been developed in cultivars of alfalfa (Sorenson et al. 1983), maize (Wiseman & Widstrom 1986, Mihm 1985), and rice (Khush 1984).

The detrimental effects of high levels of allelochemical and morphological resistance on both pest insects (insect biotypes) and beneficial insects (mortality) discussed previously demonstrate the need for the development of crop cultivars with moderate levels of insect resistance for use in IPM systems. From a cropping systems perspective, the development of individual crop cultivars with resistance to individual insect pests is of limited utility. The development of moderate levels of resistance to an insect pest complex in each of the different crops of the system is ultimately of much greater economic and ecological value to the system and its users than single crop insect resistance.

REFERENCES

Alvarado-Rodriquez, B., T. F. Leigh, K. W. Foster, and S. S. Duffey. 1986. Resistance in common bean (*Phaseolus vulgaris*) to *Lygus hesperus* (Heteroptera: Miridae). J. Econ. Entomol. 79: 484–489.

Bailey, J. C., B. W. Hanny, and W. R. Meredith, Jr. 1980. Combinations of resistant traits and insecticides: effect on cotton yield and insect populations. J. Econ. Entomol. 73: 58–60.

Beach, R. M. and J. W. Todd. 1986. Foliage consumption and larval development of parasitized and unparasitized soybean looper, *Pseudoplusia includens* [Lep.: Noctuidae], reared on a resistant soybean genotype and effects on an associated parasitoid, *Copidosoma truncatellum* [Hym.: Encyrtidae]. Entomophaga 31: 237–242.

Bell, J. V. 1978. Development and mortality in bollworms fed resistant and susceptible soybean cultivars treated with *Nomuraea rileyi* or *Bacillus thuringiensis*. J. Ga. Entomol. Soc. 13: 50–55.

Bergman, J. M. and W. M. Tingey. 1979. Aspects of interaction between plant genotypes and biological control. Bull. Entomol. Soc. Am. 25: 275–279.

Bird, L. S. 1982. Multi-adversity (diseases, insects and stresses) resistance (MAR) in cotton. Plant Dis. 66: 173–176.

Boethel, D. J. and R. D. Eikenbary (eds.) 1986. *Interactions of Plant Resistance and Parasitoids and Predators of Insects*. Ellis Horwood, Chichester, England, 224 pp.

Brattsen, L. B. 1979. Biochemical defense mechanisms in herbivores against plant allelochemics. In G. A. Rosenthal and D. H. Janzen (eds.). *Herbivores: Their Interaction with Secondary Plant Metabolites*. Academic, New York, pp. 199–270.

Brewer, G. J., E. Horber, and E. L. Sorensen. 1986. Potato leafhopper (Homoptera: Cicadellidae) antixenosis and antibiosis in *Medicago* species. J. Econ. Entomol. 79: 421–425.

Burris, E., D. F. Clower, J. E. Jones, and S. L. Anthony. 1983. Controlling boll weevil with trap cropping and resistant cotton. La. Agric. 26: 22–24.

Campbell, B. C. and S. S. Duffey. 1979. Tomatine and parasitic wasps: Potential incompatability of plant antibiosis with biological control. Science 205: 700–702.

Campbell, W. V. and J. C. Wynne. 1985. Influence of the insect-resistant peanut cultivar NC6 on performance of soil insecticides. J. Econ. Entomol. 78: 113–116.

Chalfant, R. B. 1965. Resistance of bunch bean varieties to the potato leafhopper and relationship between resistance and chemical control. J. Econ. Entomol. 58: 681–682.

Creighton, C. S., T. L. McFadden, and M. L. Robbins. 1975. Complimentary influences of host plant resistance on microbial chemical control of cabbage caterpillars. Hortic. Sci. 10: 487–488.

Creighton, C. S., T. L. McFadden, and M. L. Robbins. 1981. Comparative control of caterpillars on cabbage cultivars treated with *Bacillus thuringiensis*. J. Ga. Entomol. Soc. 16: 361–367.

Cuthbert, F. P. and R. L. Fery. 1979. Value of plant resistance for reducing cowpea curculio damage to the southernpea (*Vigna unguiculata* (L.) Walp.). J. Am. Soc. Hortic. Sci. 104: 199–201.

Cuthbert, F. P., Jr. and A. Jones. 1978. Insect resistance as an adjunct or alternative to insecticides for control of sweet potato soil insects. J. Am. Soc. Hortic. Sci. 103: 443–445.

El-Sabae, A. H., S. I. Sherby, and N. A. Mansour. 1981. Gossypol as an inducer or inhibitor in *Spodoptera littoralis*. J. Environ. Sci. Health B16: 167–178.

Felton, G. W., S. S. Duffey, P. V. Vail, H. K. Kaya, and J. Manning. 1987. Interaction of nuclear polyhedrosis virus with catechols: Potential incompatibility for host-plant resistance against Noctuid larvae. J. Chem. Ecol. 13: 947–957.

Flint, M. L. and R. van den Bosch. 1981. *Introduction to Integrated Pest Management*. Plenum, New York, 240 pp.

George, B. W. and F. D. Wilson. 1983. Pink bollworm (Lepidoptera: Gelechiidae) effects of natural infestation on upland and pima cottons untreated and treated with insecticides. J. Econ. Entomol. 76: 1152–1155.

Hamm, J. J. and B. R. Wiseman. 1986. Plant resistance and nuclear polyhedrosis virus for suppression of the fall armyworm (Lepidoptera: Noctuidae). Fla. Entomol. 69: 549–559.

Heinrichs. E. A., R. C. Saxena, and S. Chelliah. 1979. Development and implementation of insect pest management systems for rice in tropical Asia. Food and Fertilizer Technology Center Extension Bull. No. 127. 38 pp.

Heinrichs, E. A., L. T. Fabellar, R. P. Basilio, Tu-Cheng Wen, and F. Medrano. 1984. Susceptibility of rice planthoppers *Nilaparvata lugens* and *Sogatella furcifera* (Homoptera: Delphacidae) to insecticide as influenced by level of resistance in the host plant. Environ. Entomol. 13: 455–458.

Heinrichs, E. A., G. B. Aquino, S. L. Valencia, S. DeSagun, and M. B. Arceo. 1986a. Management of the brown planthopper, *Nilaparvata lugens* (Homoptera: Delphacidae), with early maturing rice cultivars. Environ. Entomol. 15: 93–95.

Heinrichs, E. A., H. R. Rapusas, G. B. Aquino, and F. Palis. 1986b. Integration of host plant resistance and insecticides in the control of *Nephotettix virescens* (Homoptera: Cicadellidae), a vector of rice tungro virus. J. Econ. Entomol. 79: 437–443.

Hudon, M., M. S. Chiang, and D. Chez. 1979. Resistance and tolerance of maize inbred lines to the European corn borer *Ostrinia nubilalis* (Hubner) and their maturity in Quebec. Phytoprotection 60: 1–22.

Isenhour, D. J. and B. R. Wiseman. 1987. Foliage consumption and development of the fall armyworm (Lepidoptera: Noctuidae) as affected by the interactions of a parasitoid, *Campoletis sonorensis* (Hymenoptera: Ichneumonidae), and resistant corn genotypes. Environ. Entomol. 16: 1181–1184.

Jenkins, J. N. 1980. The use of plant and insect models. In F. G. Maxwell and P. R. Jennings (eds.). *Breeding Plants Resistant to Insects*. Wiley, New York, pp. 215–251.

Jones, J. E., W. D. Caldwell, D. T. Bowman, J. M. Brand, A. Coco, J. G. Marshall, D. J. Boquet, R. Hutchinson, W. Aguillard, and D. F. Clower. 1981. Gumbo 500: An improved open-canopy cotton. La. Agric. Exp. Sta. Circ. 114. 14 pp.

Jones, J. E., D. James, F. E. Sistler, and S. J. Stringer. 1986. Spray penetration of cotton canopies as affected by leaf and bract isolines. La. Agric. 29: 15–17.

Kalode, M. B. 1980. The rice gall midge-varietal resistance and chemical control. In *Rice Improvement in China and Other Asian Countries*. International Rice Research Institute and Chinese Academy of Agricultural Sciences. pp. 173–193.

Kartohardjono, A. and E. A. Heinrichs. 1984. Populations of the brown planthopper, *Nilaparvata lugens* (Stal) (Homoptera: Delphaciidae), and its predators on rice varieties with differing levels of resistance. Environ. Entomol. 13: 359–365.

Kauffman, W. C. and R. V. Flanders. 1985. Effects of variably resistant soybean and lima bean cultivars on *Pediobius foveolatus* (Hymenoptera: Eulophidae), A parasitoid of the Mexican bean beetle, *Epilachna varivestis* (Coleoptera: Coccinellidae). Environ. Entomol. 14: 678–682.

Kea, W. C., S. C. Turnipseed, and G. R. Carner. 1978. Influence of resistant soybeans on the susceptibility of lepidopterous pests to insecticides. J. Econ. Entomol. 71: 58–60.

Kennedy, G. G. 1984. 2-tridecanone, tomatoes, and *Heliothis zea*: Potential incompatibility of plant antibiosis with insecticidal control. Entomol. Exp. Appl. 35: 305–311.

Kennedy, G. G., F. Gould, O. M. B. DePonti, and R. E. Stinner. 1987. Ecological, agricultural, genetic and commercial considerations in the deployment of insect resistant germplasm. Environ. Entomol. 16: 327–338.

Khush, G. S. 1984. Breeding rice for resistance to insects. Protect. Ecol. 7: 147–165.

Lampert, E. P., D. L. Haynes, A. J. Sawyer, D. P. Jokinen, S. G. Wellso, R. L. Gallun, and J. J. Roberts. 1983. Effects of regional releases of resistant wheats on the population dynamics of the cereal leaf beetle (Coleoptera: Chrysomelidae). Ann. Entomol. Soc. Am. 76: 972–980.

Leszczynski, B. 1987. Winter wheat resistance to the grain aphid *Sitobion avenae* (Fabr.) (Homoptera, Aphididae). Insect Sci. Appl. 8: 251–254.

Lewis, L. C. and R. E. Lynch. 1976. Influence on the European corn borer of *Nosema pyrausta* and resistance in maize to leaf feeding. Environ. Entomol. 5: 139–142.

Lynch, R. E. and L. C. Lewis. 1976. Influence on the European corn borer of *Nosema pyrausta* and resistance in maize to sheath-collar feeding. Environ. Entomol. 5: 143–146.

McGaughey, W. H. and R. W. Beeman. 1988. Resistance to *Bacillus thuringiensis* in colonies of Indian meal moth and almond moth (Lepidoptera: Pyralidae). J. Econ. Entomol. 81: 28–33.

McGovern, W. L. and W. H. Cross. 1976. Affects of two cotton varieties on levels of boll weevil parasitism (Coleoptera: Curculionidae). Entomophaga 21: 123–125.

Mihm, J. A. 1985. Breeding for host plant resistance to maize stem-borers. Insect Sci. Appl. 6: 369–377.

Myint, M. M., H. R. Rapusas, and E. A. Heinrichs. 1986. Integration of varietal resistance and predation for the management of *Nephottetix virescens* (Homoptera: Cicadellidae) populations on rice. Crop Prot. 5: 259–265.

Newsom, L. D., R. L. Jensen, D. C. Herzog, and J. W. Thomas. 1975. A pest management system for soybeans in Louisiana. La. Agric. 18: 10–11.

Obrycki, J. J. and M. J. Tauber. 1984. Natural enemy activity on glandular pubescent potato plants in the greenhouse: An unreliable predictor of effects in the field. Environ. Entomol. 13: 679–683.

Obrycki, J. J., M. J. Tauber, and W. M. Tingey. 1983. Predator and parasitoid interaction with aphid-resistant potatoes to reduce aphid densities: a two year field study. J. Econ. Entomol. 76: 456–462.

Orr, D. B. and D. J. Boethel. 1985. Comparative development of *Copidosoma truncatellum* (Hymenoptera: Eucyrtidae) and its host, *Pseudoplusia includens* (Lepidoptera: Noctuidae), on resistant and susceptible soybean genotypes. Environ. Entomol. 14: 612–616.

Orr, D. B. and D. J. Boethel. 1986. Influence of plant antibiosis through four trophic levels. Oecologia 70: 242–249.

Orr, D. B., D. J. Boethel, and W. A. Jones. 1985. Biology of *Telenomus chloropus* (Hymenoptera:Scelionidae) from eggs of *Nezara viridula* (Hemiptera: Pentatomidae) reared on resistant and susceptible soybean genotypes. Can. Entomol. 117: 1137–1142.

Parrott, W. L., J. N. Jenkins, and D. B. Smith. 1973. Frego bract cotton and normal bract cotton: How morphology affects control of boll weevils by insecticides. J. Econ. Entomol. 66: 222–225.

Powell, J. E. and L. Lambert. 1984. Effects of three resistant soybean genotypes on development of *Microplitis croceipes* and leaf consumption by its *Heliothis* spp. hosts. J. Agric. Entomol. 1: 169–176.

Price, P. W. 1986. Ecological aspects of host plant resistance and biological control: Interactions among three trophic levels. In D. J. Boethel and R. D. Eikenbary. (eds.). *Interactions of Plant Resistance and Parasitoids and Predators of Insects.* Ellis Horwood, Chichester, pp 11–30.

Ramoska, W. A. and T. Todd. 1985. Variation in efficacy and viability of *Beauveria bassiana* in the chinch bug (Hemiptera: Lygaeidae) as a result of feeding activity on selected host plants. Environ. Entomol. 14: 146–148.

Reissig, W. H., E. A. Heinrichs, L. Antonio, M. M. Salac, A. C. Santiago, and A. M. Tenorio, 1981. Management of pest insects of rice in farmers fields in the Philippines. Prot. Ecol. 3: 203–218.

Reynolds, G. W., C. M. Smith, and K. M. Kester, 1984. Reductions in consumption, utilization and growth rate of soybean looper (Lepidoptera: Noctuidae) larvae fed foliage of soybean genotype PI227687. J. Econ. Entomol. 77: 1371–1375.

Robinson, J. F., E. C. Berry, L. C. Lewis, and R. E. Lynch. 1978. European corn borer: Host-plant resistance and use of insecticides. J. Econ. Entomol. 71: 109–110.

Rose, R. L., T. C. Sparks, and C. M. Smith. 1988. Insecticide toxicity to larvae of *Pseudoplusia includens* (Walker) and *Anticarsia gemmatalis* (Hubner) (Lepidoptera) as influenced by feeding on resistant soybean (PI 227687) leaves and coumestrol. J. Econ. Entomol. 81: 1288–1294.

Salto, C. E., R. D. Eikenbary, and K. J. Starks. 1983. Compatibility of *Lysiphlebus testaceipes* (Hymenoptera: Braconidae) with greenbug (Homoptera: Aphididae) biotypes "C" and "E" reared on susceptible and resistant oat varieties. Environ. Entomol. 12: 603–604.

Saxena, R. C. 1982. Colonization of rice fields by *Nilaparvata lugens* (Stal) and its control using a trap crop. Crop Prot. 1: 191–198.

Schuster, M. F. and M. Calderon. 1986. Interactions of host plant resistant genotypes and beneficial insects in cotton ecosystems. In D. J. Boethel and R. D. Eikenbary (eds.). *Interactions of Plant Resistance and Parasitoids and Predators of Insects.* Ellis Horwood, Chichester, pp. 84–97.

Schuster, M. F., M. J. Lukefahr, and F. G. Maxwell. 1976. Impact of nectariless cotton on plant bugs and natural enemies. J. Econ. Entomol. 69: 400–402.

Sorenson, E. L., E. K. Horber, and D. L. Stuteville. 1983. Registration of KS 80 alfalfa germplasm resistant to the blue alfalfa aphid, pea aphid, spotted alfalfa aphid, anthracnose, and downy mildew. (Reg. No. GP 126). Crop Sci. 23: 599.

Starks, K. J., J. R. Muniappan, and R. D. Eikenbary. 1972. Interaction between plant resistance and parasitism against greenbug on barley and sorghum. Ann. Entomol. Soc. Am. 65: 650–655.

Teetes, G. L. 1980. Breeding sorghums resistant to insects. In F. G. Maxwell and P. R. Jennings (eds.). *Breeding Plants Resistant to Insects*. Wiley, New York, pp. 459–485.

Teetes, G. L., M. I. Becerra, and G. C. Peterson. 1986. Sorghum midge (Diptera: Cecidomyiidae) management with resistant sorghum and insecticide. J. Econ. Entomol. 79: 1091–1095.

Treacy, M. F., G. R. Zummo, and J. H. Benedict. 1985. Interactions of host-plant resistance in cotton with predators and parasites. Agric. Ecosyst. Environ. 13: 151–158.

Treacy, M. F., J. H. Benedict, M. W. Walmsley, J. D. Lopez, and R. K. Morrison, 1987. Parasitism of bollworm (Lepidoptera: Noctuidae) eggs on nectaried and nectariless cotton. Environ. Entomol. 16: 420–423.

van Emden, H. F. 1986. The interaction of plant resistance and natural enemies: Effects on populations of sucking insects. In D. J. Boethel and R. D. Eikenbary (eds.). *Interactions of Plant Resistance and Parasitoids and Predators of Insects*. Ellis Horwood, Chichester, pp. 138–150.

Van Haren, R. J. F., M. M. Steenhuis, M. W. Sabelis, and O. M. B. dePonti. 1987. Tomato stem trichomes and dispersal success of *Phytoseiulus persimilis* relative to its prey, *Tetranychus urticae*. Exp. Appl. Acarol. 3: 115–121.

Wiseman, B. R. and N. W. Widstrom. 1986. Mechanisms of resistance in 'Zapalote Chico' corn silks to fall armyworm (Lepidoptera: Noctuidae) larvae. J. Econ. Entomol. 79: 1390–1393.

Wiseman, B. R., W. W. McMillian, and N. W. Widstrom. 1972. Tolerance as a mechanism of resistance in corn to the corn earworm. J. Econ. Entomol. 65: 835–837.

Wiseman, B. R., E. A. Harrell, and W. W. Mcmillian. 1975. Continuation of tests of resistant sweet corn hybrid plus insecticides to reduce losses from corn earworm. Environ. Entomol. 2: 919–920.

Wiseman, B. R., W. W. Mcmillian, and N. W. Widstrom. 1976. Feeding of corn earworm in the laboratory on excised silks of selected corn entries with notes on *Orius insidiosus*. Fla. Entomol. 59: 305–308.

Yanes, J., Jr. and D. J. Boethel. 1983. Effect of a resistant soybean genotype on the development of the soybean looper (Lepidoptera: Noctuidae) and an introduced parasitoid, *Microplitis demolitor* Wilkinson (Hymenoptera: Braconidae) Environ. Entomol. 12: 1270–1274.

Yu, S. J. 1982. Induction of microsomal oxidases by host plants in the fall armyworm, *Spodoptera frugiperda* (J. E. Smith). Pestic. Biochem. Physiol. 17: 59–67.

Yu, S. J. 1983. Induction of detoxifying enzymes by allelochemicals and host plants in the fall armyworm. Pestic. Biochem. Physiol. 19: 330–336.

Yu, S. J. and E. L. Hsu. 1985. Induction of hydrolases by allelochemicals and host plants in fall armyworm (Lepidoptera: Noctuidae) larvae. Environ. Entomol. 512–515.

Index